I Am a Man of Peace

Writings Inspired by the Maynooth University Ken Saro-Wiwa Collection

I Am a Man of Peace

Writings Inspired by the Maynooth University Ken Saro-Wiwa Collection

Edited by Helen Fallon

Daraja Press

Published by
Daraja Press
https://darajapress.com

Maynooth University is not responsible for the opinions or views expressed in this book.

Library and Archives Canada Cataloguing in Publication

Title: I am a man of peace : writings inspired by the Maynooth University Ken Saro-Wiwa Collection
/ edited by Helen Fallon.
Names: Fallon, Helen, editor. | Ken Saro-Wiwa Poetry Competition.
Description: Contains essays by a wide variety of international contributors as well as poems from the Ken Saro-Wiwa Poetry Competition. | Includes bibliographical references.
Identifiers: Canadiana (print) 20200289772 | Canadiana (ebook) 2020029024X | ISBN 9781988832708
(softcover) | ISBN 9781988832715 (EPUB)
Subjects: LCSH: Saro-Wiwa, Ken, 1941-1995. | LCSH: Ken Saro-Wiwa Collection (Maynooth University) |
LCSH: Social justice. | LCSH: Human rights. | LCSH: Environmental protection. | LCSH: Social justice–Poetry. | LCSH: Environmental protection–Poetry.
Classification: LCC HM671 .I33 2020 | DDC 303.3/72–dc23

Cover design by Leo Duffy of Yellowstone

Contents

Section 1 Essays

This book is dedicated to the memory of the Ogoni 9

Baribor Bera
Saturday Dobee
Nordu Eawo
Daniel Gbokoo
Barinem Kiobel
John Kpuinen
Paul Levura
Felix Nuate
Ken Saro-Wiwa

Foreword

On 10 November 2011, the 16th anniversary of the execution of the Ogoni 9, Maynooth University President, Professor Philip Nolan accepted, on behalf of the University, the donation from Sister Majella McCarron, of the Ken Saro-Wiwa archive. Sister Majella's choice of Maynooth University for this unique donation was particularly appropriate, given the University's long involvement with issues of inclusion and justice in Ireland and abroad. This deep-rooted commitment is today articulated in our University Strategic Plan, where a strategic goal is *'to build on our achievements to date and become a model University for equality, diversity, inclusion and inter-culturalism, where social justice, addressing inequality and empowering people are central to our mission'*, and through the establishment of a Vice-President for Equality and Diversity and a dedicated Equality and Diversity office in the University.

The letters and poems, written by Ken Saro-Wiwa between 20 October 1993 and 14 September 1995, were smuggled out of military detention in breadbaskets. Their journey from the Niger Delta to Ireland is, in itself, a fascinating aspect of the story. The letters shed light on Saro-Wiwa as a writer, an activist and a family man who was executed, with eight colleagues (The Ogoni 9), for protesting against the activities of the international petrochemical industry in their homeland Ogoni, in the Niger Delta. The collection casts a unique light on a complex conflict over ownership of natural resources and environmental destruction.

Maynooth University has a dedicated team of librarians, archivists and conservators and expertise in digitisation and open access. Much work has been done to preserve, conserve and make available both physically and digitally, this major collection. The letters and poems have been published as *Silence Would be Treason: Last Writings of Ken Saro-Wiwa* (2013, 2018)[1] and, in keeping with the Library's commitment to open access, and the desire to have the material available to activists and researchers in the Global South, the book is freely available via the Maynooth University Institutional Repository.

We do not have the letters Sister Majella sent to Saro-Wiwa, which, as she explains in her essay, are likely to have been destroyed. However, the Ken Saro-Wiwa Audio Archive, created by Kairos Communications and Maynooth University Library, has extensive recordings of her life story and experiences in Nigeria and Ireland working with Saro-Wiwa to highlight the plight of the

1 Corley, I. Fallon, H. & Cox, L. (2018) *Silence Would be Treason: Last Writings of Ken Saro-Wiwa.* Senegal/Montreal: Daraja Press. 2nd edition. http://mural.maynoothuniversity.ie/10161/

Ogoni people. The audio archive also contains recordings of people connected with Saro-Wiwa, including his brother Dr Owens Wiwa and his daughter, renowned travel writer Noo Saro-Wiwa. Maynooth University has had the pleasure of hosting both on visits to speak at the annual Ken Saro-Wiwa Seminar and to view the collection. As with *Silence Would be Treason: Last Writings of Ken Saro-Wiwa* the audio archive is available on open access.[2]

The Ken Saro-Wiwa Archive is probably one the Library's most hard-working collections. The letters have been exhibited to mark events such as International Human Rights Day, Africa Day, Development Studies Week, the anniversary of the execution of the Ogoni Nine (November 10th 1995) and to coincide with conferences/seminars which have a development studies/conflict resolution theme. Schoolchildren have visited the Library, viewed the letters and discussed the issues surrounding the conflict, and the broader issue of climate justice and land rights, in class. Students in Transition Year and above, have had the opportunity to take part in the Ken Saro-Wiwa poetry competition and those honoured are featured among the 41 poems, by new and established poets in this collection. Material from the archive has been integrated into elements of the postgraduate and undergraduate curriculum, an initiative described in one of the essays.

The first international exhibition of material from the collection opened at Quinnipiac University in Connecticut in 2016. A Ken Saro-Wiwa travelling exhibition in Irish public libraries commenced in January 2019. Work with an American filmmaker, who hopes to make a film on the Saro-Wiwa story, has also commenced.

Locally and globally, justice and equality work continues. The 21 essays in this collection explore key global issues including climate change, environmental pollution, the rights of minority groups in the face of abuse by large corporations, and promoting diversity and equality in all its aspects. The death-row correspondence, has been the catalyst for this collection, which I hope, will inspire us to continue the justice and equality quest of Ken Saro-Wiwa.

Gemma Irvine
Vice-President Equality & Diversity
Maynooth University

October 2020

2 Ken Saro-Wiwa Audio Archive: https://www.maynoothuniversity.ie/library/collections/ken-saro-wiwa-audio-archive

Introduction

I feel privileged to bring together this collection of essays and poems to mark the 25th anniversary of the execution of Nigerian activist and writer Ken Saro-Wiwa and his eight colleagues (the Ogoni 9). The 21 essays and 42 poems included here are inspired by his ideals and activism.

In 2011, John O'Shea, then a postgraduate student in the Sociology Department at Maynooth University (MU), approached the Library regarding a possible donation of the death row correspondence from Saro-Wiwa to Sister Majella McCarron (OLA). She had worked with Saro-Wiwa in Nigeria to highlight the environmental destruction of Ogoniland by Royal Dutch Shell. Later, back in Ireland, she campaigned nationally and internationally to save the lives of the Ogoni 9.

John had interviewed Sister Majella when investigating the topic of media coverage of the Shell to Sea Campaign for his thesis. One of her two essays in this volume recounts her experience as a table observer of the Shell to Sea campaign, which strove to have gas, discovered off the west coast of Ireland, refined at sea rather than inland.

I had met Sister Majella a few years previously when she was undertaking postgraduate studies at Mary Immaculate College in Limerick. She was aware I had spent time teaching at the University of Sierra Leone and called to the Library to get my assistance in sourcing information for a project. She had taught at the University of Lagos, and we had an interesting exchange of experience. Both of us had seen the effects of corrupt and inept government and the power of large multinational corporations. There were similarities between Sierra Leone and Nigeria. The Niger Delta, where Ogoni is located, is rich in oil: Sierra Leone had an abundance of diamonds, gold and other resources. Rather than being of benefit to the people, the mismanagement of these resources and other complex historical and political factors, resulted in a ten-year civil war in the case of Sierra Leone, and in the case of Ogoni, environmental devastation, the destruction of people's livelihoods, and the deaths of the Ogoni 9.

A few years after that first encounter with Sister Majella, John O'Shea approached the Library. Recognising the value of the collection, the Library immediately began working with Sister Majella and the Saro-Wiwa family to preserve and make available Ken's legacy. The cataloguing, digitisation and subsequent use of the archive to support teaching and research are the subject of an essay by MU Library archivists Ciara Joyce and Roisin Berry, while Hugh Murphy, Head of Collections and Content, looks at the broader, often complex, issues of collection development in his essay.

In 2013, the book *Silence Would be Treason: Last Writings of Ken Saro-Wiwa*, containing the death row correspondence, poems and contextual essays, was published by Daraja Press, with a second edition published in 2018.[1] Dr Owens Wiwa, brother of Ken Saro-Wiwa, visited the Library and launched the book. In his essay, the first in this volume, he recounts how his older brother awakened and nurtured his awareness of the tremendous damage being wrought by Royal Dutch Shell to their homeland, in collaboration with the then Nigerian military dictatorship. His firsthand account of the brutality of the military government and its impact; his unsuccessful efforts to save the life of his brother; going into hiding and subsequently escaping with his family from Nigeria; and his efforts to retrieve the remains of his brother for burial, make for a very moving read.

In 2015, Noo Saro-Wiwa visited MU Library to view the archive and read from her award-winning travel book *Looking for Transwonderland: Travels in Nigeria*.[2] A recording of the reading is available on the library YouTube channel.[3] In the second essay of this collection, she shares her story of growing up in England with strong links to family in Nigeria, and the trauma of hearing of her father's execution while at university.

Sister Majella McCarron provides two personal essays. The first is a reflection on the events that shaped her work with Saro-Wiwa in Nigeria and her subsequent efforts to save the lives of the Ogoni 9. The second recounts her work on the Shell to Sea and other campaigns in Ireland.

The environmental destruction that Shell has caused in the Niger Delta is addressed by Mark Dummett, Head of Business, Security and Human Rights at Amnesty International. This organisation investigated and documented how Shell and other oil companies have caused or contributed to human rights abuses through their operations in the Niger Delta. Daniel Leader, a barrister and partner at Leigh Day's international law department, known for leading a number of ground-breaking human rights cases, including a series of cases against Shell on behalf of Nigerian communities, explores the issue of legal redress. His essay recounts how a new

1 Corley, I., Fallon, H. & Cox, L. (2018). *Silence Would be Treason: Last Writings of Ken Saro-Wiwa*. Senegal: Daraja Press. http://mural.maynoothuniversity.ie/10161/

2 Saro-Wiwa, Noo (2012) *Looking for Transwonderland: Travels in Nigeria*. London: Granta

3 Noo Saro-Wiwa reading at Maynooth University Library https://www.youtube.com/watch?v=ZHt49EeHsgk&t=356s

generation of activists and lawyers are seeking to hold Shell and other multinationals to account for the "ecological war" they have waged in the Niger Delta.

The United Nations Environmental Programme (UNEP) report of the Environmental Assessment of Ogoniland,[4] recorded that drinking water in Ogoni had benzene, a carcinogen, at over 900 times the level permitted. Nigerian architect, environmental activist, author and poet Nnimmo Bassey's wide ranging essay draws on this report and Saro-Wiwa's writings in discussing Saro-Wiwa as activist, writer and creator of the Ogoni Bill of Rights. Anthropologist Dr Abayomi Ogunsanya explores the Niger Delta cultural landscape in his essay, while Dr Samuel Udogbo's essay draws on his MU doctoral research which examines Ogoni's resistance in Nigeria. Firoze Manji examines the commonalities between Amilcar Cabral, the Guinea-Bissau revolutionary, and Ken Saro-Wiwa, focusing on the centrality of culture in the search for freedom.

Dr Laurence Cox discusses what we can learn from how the Ogoni, a small rural group, remote from the centres of power, were able to effectively resist Shell. He argues that it is only by challenging the priorities of those who hold power and pushing for a radically different kind of economy, with the attendant changes in society, politics and culture, that we can hope to avert climate disaster.

The final essays, while inspired by Saro-Wiwa's quest for equality and justice, reflect on aspects of an increasingly diverse Irish society. Veronica Akinborewa, Dr Camilla Fitzsimons and Philomena Obasi use an auto ethnographic approach – a conversation between three adult educators – in their essay, to explore relationships that exist amidst the intersections of race, gender and institutional positions. The three have designed and delivered workshops, with MU Department of Adult and Community Education, on culture, interculturalism and racism. One such workshop is the topic of the essay on diversity training for library staff at MU by Helen Fallon, Laura Connaughton and Edel Cosgrave. The training was part of the Library Strategic Plan 2020-2023, which has Equality, Diversity, Inclusion and Interculturalism as a strategic goal. Maynooth University's commitment to inclusion is also articulated in Dr Cliodhna Murphy's essay on the University of Sanctuary.

4 UNEP (2011) Environmental Assessment of Ogoniland Report. https://www.unenvironment.org/explore-topics/disasters-conflicts/where-we-work/nigeria/environmental-assessment-ogoniland-report

The second section of the book contains poems by both established and new poets, including contributions from schoolchildren in their senior cycle, who have visited the University and viewed the collection. The poems are preceded with a contextual essay by Irish poet and creative writing teacher Jessica Traynor, who has worked with Maynooth University Library on delivering poetry workshops for schoolchildren and adults, both face-to-face and, more recently, virtually via Zoom. While the workshops grew from human rights violations in Nigeria, they sought to inspire people to write about their own life experiences. Jessica Traynor's essay, which opens this section, provides additional insights into this aspect of the collection. The concluding essay is by David Rinehart from MU Library, who has worked with migrant aid and solidarity organisations for many years. He reflects on the poems in the collections for both their intrinsic beauty and as a tool for looking both inward and outward in order to better understand different ways of being and the effect of our actions as a global community on the people we share this planet with.

Writing to Sister Majella McCarron on 1 December 1993, Saro-Wiwa urged her:

> *Keep putting your thoughts on paper. Who knows how we can use them in future. The Ogoni story will have to be told.*[5]

I hope this rich and diverse collection of essays and poems goes some way to fulfilling his wishes, and strengthens in us – writers, poets and readers – the resolve to tell the stories of marginalised and oppressed communities everywhere. Our voices can help create a world where justice and equality are the cornerstone of our societies.

Helen Fallon
Deputy University Librarian

5 Maynooth University Ken Saro-Wiwa Archive PP7/2

Acknowledgements

Many people helped bring this book to birth. I am grateful to Sister Majella McCarron (OLA) for her commitment and enthusiasm for the project. Particular thanks are due to the Saro-Wiwa family, especially Owens Wiwa and Noo Saro-Wiwa for their engagement. To all the writers and poets in the collection, my sincere thanks.

Many of the Maynooth University Library staff helped in a number of ways. I am grateful to these wonderful colleagues who continue to inspire me by their commitment to the creation and provision of information, under the leadership of University Librarian Cathal McCauley, who has supported this project from the outset.

Firoze Manji, from Daraja Press, a not-for-profit publisher, which has the aim of nurturing reflection, sheltering hope and inspiring audacity, has been as always, wonderful to work with. Thanks to Leo Duffy of Yellowstone for producing a beautiful cover.

Finally, thanks to my partner Joe and all my family for their ongoing support in all my endeavourss.

About the Contributors

Yewande Adebowale is a Nigerian lawyer, storyteller, poet and author of two collections of poems titled *A tale of being, of green and of ing.* (2019) and *Voices: A collection of poems that tell stories* (2016). Her poems have appeared in *Visual Verse, Afritondo, Trampset, Poemify, Pride Magazine, Lumiere Review* and elsewhere.

Veronica Akinborewa is a Nigerian-born Irish citizen. Her passion is for working with marginalized communities. She has been involved in a number of projects including a choral choir group, in which she held a voluntary co-ordinating position. She was also a volunteer with substance misusers and with marginalized women in an area of high deprivation She currently works with a housing organisation and is also a placement tutor on an initial teacher training course in Maynooth University, where she was a postgraduate student.

Elizabeth Akinwande attended Donabate Community College. She is now a first year student at University College Dublin (UCD) studying social justice, politics and international relations. She has written for *Fighting Words* and *The Irish Times* No Child 2020 project. She plans to keep writing, and to "help people with the things I'm writing."

Ceri Arnott completed her secondary school education at Wesley College Dublin in 2020. There she was actively involved in the school's Amnesty International group. She is now studying medicine at UCD. She enjoys spending her spare time doing gymnastics and reading. She plans to stay involved in activist organisations that seek a better world for all and will continue writing for leisure.

Nnimmo Bassey is director of the ecological think-tank Health of Mother Earth Foundation (HOMEF) based in Nigeria. He chaired Friends of the Earth International from 2008 to 2012. His books include *To Cook a Continent – Destructive Extraction and the Climate Crisis in Africa* (Pambazuka Press, 2012) and *Oil Politics – Echoes of Ecological War* (Daraja Press, 2016).

Roisin Berry has a background in archaeology, art history and archival studies. Her work has involved preserving and promoting the archives of local authorities, banks, and academic institutions, including Ulster Bank, University of Limerick, Clare County Council, and Royal Irish Academy. She has been a member of the Special Collections and Archives team at Maynooth University Library since 2008. With a special interest in literary archives, she was responsible for cataloguing the Ken Saro-Wiwa Archive.

Gavin Bourke grew up in Tallaght, Dublin. He holds a B.A. Degree in Humanities from Dublin City University, an M.A. in Modern Drama Studies from University College Dublin and a Higher Diploma in Information Studies. His work covers nature, time, memory, addiction, mental health, human relationships, inner and outer life, creating meaning and purpose, politics, contemporary and historical social issues, injustice, the human situation, power and its abuse, as well as urban and rural life. He has been published widely.

Caroline Bracken's work has been published in *The Irish Times, The Fish Anthology, Sonder Magazine, The Ogham Stone* and the forthcoming *Best New British and Irish Poets* (Eyewear UK). She was selected for the Poetry Ireland Introductions Series 2018 and has been sponsored by Culture Ireland to read her work in the USA. She recently won the Poetry Day Ireland 2020 and the DLR/Creative Ireland Poetry competitions. One of her poems features in *The Poetry Jukebox*.

David Butler is a multi-award-winning novelist, poet, short-story writer and playwright. His third novel, *City of Dis* (New Island) was shortlisted for the Kerry Group Irish Novel of the Year, 2015. His second poetry collection, *All the Barbaric Glass*, was published in 2017 by Doire Press. Arlen House is to bring out his second short story collection *Fugitive*, in 2020. Literary prizes for poetry include the Maria Edgeworth, Féile Filíochta, Ted McNulty, Brendan Kennelly, Baileborough and Poetry Ireland/Trócaire awards.

My name is **Maeve Byrnes**. I'm 17 years old. I am in my final year of school at Maynooth Community College. Sport is my biggest passion, therefore I would love to study sport in college. I have an interest in physiotherapy, taking care of sports injuries or physical rehabilitation.

Laura Connaughton is Head of Academic Services in Maynooth University Library. She has a Master's degree in Library and Information Studies from University College Dublin and a B.A. in Languages with Computing from University of Limerick. She has presented at national and international conferences and published in the area of library service provision. She was the recipient of the 2016 CONUL (Consortium of National and University Libraries) Conference Poster Award and presents workshops on designing posters regularly.

Edel Cosgrave joined Maynooth University in 2008 as a Library Assistant. She is a member of the Library Engagement and Information Services team whose main focus is excellence in the delivery of front-line Library services. She holds a B.A. from National University of Ireland Galway (NUIG) and prior to joining the Library worked in the field of IT service delivery for several different organisations.

Laurence Cox is Associate Professor of Sociology at Maynooth University and co-editor, with Helen Fallon and Íde Corley, of *Silence Would be Treason: Last Writings of Ken Saro-Wiwa*. He has a long-standing involvement in ecological, international solidarity and other movements and has written and published widely on the subject, including *Why Social Movements Matter* and *The Irish Buddhist: the Forgotten Monk who Faced Down the British Empire*.

Siobhan Curran is the Advocacy and Policy Advisor on Human Rights and Democratic Space with Trócaire. She previously coordinated the Roma project in Pavee Point Traveller and Roma Centre and has worked in the areas of human rights and gender, including with Amnesty International Ireland. She has a Master's degree in Human Rights Law and Transitional Justice from Ulster University and a Master's degree in Social Policy from University College Dublin.

Tony Daly is project manager of online global issues platform developmenteducation.ie, and co-ordinator of 80:20 Educating and Acting for a Better World which promotes popular education on human development and human rights. He is editor of the development issues primer *80-20 Development in an Unequal World*, 7th Edition, published by the New Internationalist. He is also a member of the Women's Rights and Gender Equality Working Group (GWG) of EuroMed Rights and a trustee of Fairtrade Ireland.

Marykate Donohoe attends Presentation Secondary School, Wexford. In March 2020 the school awarded her Senior Writer of the Year for her poetry. She is an enthusiastic member of her school's writing register and journalism class. Marykate is from Rosslare and a past pupil of Scoil Mhuire where her 2016 updated version of the Irish Proclamation is proudly displayed.

Áine Dooley is a Leaving Certificate Student at Maynooth Community College. She is a keen sportsperson with a particular passion for Gaelic football and soccer. She enjoys English literature and drama.

Mark Dummett is Head of Business, Security and Human Rights at Amnesty International. Previously he worked with the BBC. Since 2015 he has investigated how Shell and other oil companies have caused or contributed to human rights abuses through their operations in the Niger Delta.

Chiamaka Enyi-Amadi is a writer, performer, arts facilitator, and literary editor. Her work is published in *Poetry International 25, Poetry Ireland Review 129, IMMA Magazine, Architecture Ireland, The Irish Times, Writing Home: The New Irish Poets anthology* (Dedalus Press 2019, co-edited by Pat Boran & Chiamaka Enyi-Amadi) and *The Art of the Glimpse: 100 Irish Short Stories* (Head of Zeus 2020, edited by Sinead Gleeson). She has also been featured on the RTÉ Poetry Programme.

Helen Fallon is Deputy Librarian at Maynooth University. She has previously worked at Dublin City University and the University of Sierra Leone. She has published extensively and presents workshops on writing for academic publication nationally and internationally. She edited (with Íde Corley and Laurence Cox) the death row correspondence of Ken Saro-Wiwa, published as *Silence Would be Treason: Last Writings of Ken Saro-Wiwa*. She created the Maynooth University Ken Saro-Wiwa Audio Archive with Dr Anne O'Brien.

Eilish Fisher grew up on a farm in rural Vermont and moved to Ireland in 1998. Her poetry has been published in magazines and literary journals: most recently in the anthology *Writing Home*, published by Dedalus Press in Autumn 2019. Her first children's novel, set in the mountains and valleys of County Wicklow, was sort-listed for the Mslexia Children's Novel Award. She has a Master's degree in Early Medieval Irish History and Literature and a Doctorate in Medieval English Literature from Maynooth University.

Dr Camilla Fitzsimons is a Lecturer in Adult and Community Education at Maynooth University. She currently manages the Higher Education in Further Education and initial teacher education programme for Further Education practitioners. Previously she worked in leadership roles in civil society community education organisations. She is interested in critical pedagogy, feminism, teaching practice, race and racism, equality and social justice. She has published extensively on these topics at all times seeking to uncover dimensions of power that are often unnamed.

Bairbre Flood is a writer and journalist with a special interest in migration/refugee stories. She has a podcast, Short Talks, and has had several radio documentaries broadcast on Newstalk and community stations. She won the Fish Memoir Competition in 2019, has pieces published in *The Cormorant* and *Hinterland Magazine,* and is currently working on a video poem series, *In This Together?* with a photographer in Moria Refugee Camp. Contact her on twitter: @bairbreflood

Lind Grant-Oyeye was born in the Niger Delta and now lives in Kilkenny. She is passionate about highlighting social justice and mental health issues through the arts. When she is not writing, she is engaged in advocacy for mental health services for young people. She is the recipient of various poetry awards and has work published in literary magazines, anthologies and curated poetry projects around the world.

Mary Melvin Geoghegan has published extensively including five collections of poetry. Her most recent collection is *As Moon and Mother Collide* published by Salmon Poetry in 2018. She has won a number of national poetry competitions and is a member of the Writers in School Scheme with Poetry Ireland and has edited several anthologies of children's poetry.

Dr Gemma Irvine is Vice-President of Equality & Diversity in Maynooth University, leading strategic change in the areas of EDI for both staff and students. Previously Head of Policy and Strategic Planning in the Irish Higher Education Authority (HEA), Dr Irvine coordinated a national higher education policy approach with specific responsibility for: enhancement of teaching and learning; research including open science; international education; and EDI. Dr Irvine completed her PhD in Neuroscience in New Zealand before coming to Ireland in 2004.

Ciara Joyce is an archivist in Maynooth University Library. She is a graduate of UCD's Higher Diploma in Archival Studies and has extensive experience caring for both public and private collections. Her work has focused on collection advocacy including through outreach and education projects. She has curated numerous exhibitions including a major permanent exhibition at Airfield, Dublin and a substantial travelling exhibition to mark the fourth centenary of the Flight of the Earls, and is currently listing the papers of writer and poet Pearse Hutchinson.

Daniel Leader is a Barrister and Partner at Leigh Day's international law department. He specialises in international human rights and environmental law, with a particular focus on business and human rights. Over the past 25 years Leigh Day has brought ground breaking cases against parent companies on behalf of victims from the developing world who have sought redress for human rights abuses committed by British companies, including a series of cases against Shell on behalf of Nigerian communities who have suffered from systemic pollution.

Firoze Manji is the publisher of Daraja Press and Adjunct Professor in the Institute for African Studies at Carleton University, Ottawa, Canada. He is the founder and former editor of *Pambazuka News* and publisher of Pambazuka Press. He has published widely on health, human rights, development and politics and has edited several books including *African Awakenings: The Emerging Revolutions* and *Claim No Easy Victories: The Legacy of Amilcar Cabral.*

Gillian Muir - My life includes having taught *Paradise Lost* in post-civil war Nigeria, surviving in London and, in 2000, settling into six years on a remote Scottish island. I record all my experiences through my art works which are acrylic on canvas, cloth wall-hangings made from refreshed textiles, and by poetry. By definition, I am a grandmother who writes, sews and paints.

Clíodhna Murphy is an Associate Professor at Maynooth University Department of Law. She lectures and researches on the topic of migration and human rights. She is currently chair of the Maynooth University Sanctuary Steering Committee and represents Maynooth on the national Universities and Colleges of Sanctuary Ireland committee.

Hugh Murphy is Head of Collections and Content at Maynooth University Library. He has worked previously in University College Dublin (UCD) Library and in the National Library of Ireland as well as lecturing in Library Studies in UCD and book history and archival studies in Maynooth University. He is currently pursuing doctoral studies in early 19th-century history. His main professional interests lie in the areas of collection development and resource description and he has published and spoken nationally and internationally on these topics.

My name is **Owodunni Ola Mustapha**, I am a Nigerian, a graduate of Political Science and a mum of 3. I am an asylum seeker, an activist, a writer and aspiring poet. I have had some of my work published in journals in Ireland including *Up the hill in Mayo*, published in *Correspondences: an Anthology* to call for an end to direct provision, and *The Unknown*, published in *MASI journal* in 2019. I am the founder of Ballyhaunis Inclusion Project and in 2019 I was honored with the Christine Buckley Volunteer of the year award.

Paul McCarrick's work has been published in *Poetry Ireland Review, The Stinging Fly, The Ogham Stone*, and elsewhere. He was selected for the Poetry Ireland Introductions Series 2019 and received an Artist Bursary Grant from Westmeath Arts Office. He is currently living in the midlands in Ireland where he is completing his first collection.

Majella McCarron grew up in Derrylin in Fermanagh. She joined the Missionary Institute of Our Lady of Apostles in 1956. After graduating with a science degree from University College Cork, she taught for 30 years in Nigeria, at secondary school level and at the University of Lagos. She worked closely with Ken Saro-Wiwa on issues of justice and the environment and was compelled to campaign for the lives of the Ogoni 9, who were hanged on the 10th of November, 1995. In 2011, she donated the death-row correspondence she received from Ken Saro-Wiwa, to Maynooth University Library.

Nora Nadjarian is an award-winning poet and writer from Cyprus. She has had poetry and short fiction published internationally. Her work was included in *A River of Stories*, an anthology of tales and poems from across the Commonwealth, *Best European Fiction 2011* (Dalkey Archive Press), *Being Human* (Bloodaxe Books, 2011), *Capitals* (Bloomsbury, 2017) and, most recently, *Europa 2* (Comma Press, 2020). Her latest book is the collection of short stories *Selfie* (Roman Books, 2017).

Philomena Ilobekeme Obasi works in social care and adult and community education. She facilitates adult and youth groups, addressing the changing trends of Irish education which embraces diversity and interculturalism. She works with children and youth from different countries through leadership and mentoring programmes and is currently a volunteer as an Administrator of a youth group. Philomena is an occasional lecturer with Maynooth University, and lives with her family in Kildare. Her interests include reading, writing, music and culture.

Christeen Udokamma Obasi moved with her family from Nigeria to Ireland in 2014. She completed her secondary school education at Maynooth Community College and is now a first year student at Trinity College Dublin, studying Biological and Biomedical Sciences. Her hobbies include writing, dancing, music, cooking, swimming and photography. She is a member of the National Youth Council of Ireland with interests in youth global affairs, international current issues, climate change, justice and equity.

Abayomi Ogunsanya is an anthropologist and a writer. He holds a PhD degree in Anthropology from the University of Ibadan, Nigeria, where he spent two years teaching Anthropology in the Institute of African Studies. In 2018 he moved to Ireland with his family and currently lives in Galway.

Dr Anne O'Brien is a Lecturer with the Department of Media Studies at Maynooth University. She has published a number of articles on the representation of women in radio and television and on women workers in creative industries, and examined why women leave careers in screen production. She has also undertaken research on community media in funded projects for the Broadcasting Authority of Ireland. Her most recent book is *Women, Inequality and Media Work* (Routledge, 2019).

Jimmy O'Connell was born in Dublin. He is a graduate of University College Dublin (UCD). He has been writing and performing his work for many years in the Irish Writers Centre, Sunflower Sessions and other venues. His poetry has appeared in *The Baltimore Review, Poetry Ireland Review, Stepaway Magazine, Flare 7, Poetry for a New Ulster.* He has had a collection of his poetry *Although it is Night* published by Wordonthestreet in 2013. In 2018 he published his first novel *Batter the Heart.*

Liam O'Neill is a poet, writer and activist living in Galway City. His work has been published in *The Irish Times, Poetry Ireland Review, The Children of the Nation Anthology,* and *Extinction Rebellion Creative Hub.* He was a winner in the Strokestown 20/20 prize, was shortlisted for the Wexford Poetry Competition and longlisted for The Moth Poetry Prize in 2020. Much of Liam's writing deals with social and environmental issues.

Patrick O'Siochru was born in Russia and raised in Ireland. He studied Theology & German at Maynooth University and is interested in writing in different forms. He is also interested in sports and music.

Eva Paturyan – I am a sixth year student at Maynooth Community College where English is one of my favourite subjects. I have been an avid reader for as long as I can remember and have always enjoyed writing anything from short stories to history essays. I enjoy learning about the stories of the many countries I have visited. I have been playing the violin from a very young age and I plan to continue music studies into third level and hope to have a career in music in the future.

Ciara Regan is education and research officer at 80:20 Educating and Acting for a Better World, staff writer at developmenteducation.ie and project leader on the longitudinal audit of Irish development education resources research study. Ciara has researched and published in the area of women and development in the context of HIV and AIDS in Zambia. She is also editor of the development issues primer *80-20 Development in an Unequal World,* 7th Edition, published by the New Internationalist and 80:20.

David Rinehart grew up in Florida where he was raised by his American father and Venezuelan mother. He moved to Ireland with his partner and daughter in 2018. He received an M.A. in Latin American Studies in 2018. David has worked with migrant aid and solidarity organisations for many years and continues to fight for a just and equal world. David is a Library Assistant at Maynooth University Library's Special Collections and Archives Department and is undertaking a Master's degree in Library and Information Science from Robert Gordon University.

Noo Saro-Wiwa is a travel writer and author. Her first book, *Looking for Transwonderland: Travels in Nigeria* (Granta, 2012), was selected as BBC Radio 4's Book of the Week in 2012, and named *The Sunday Times* Travel Book of the Year. In 2016 *Looking for Transwonderland* won the Albatros Travel Literature Prize in Italy. It has been translated into French and Italian. She is the daughter of Ken Saro Wiwa.

My name is **Zofia Terzyk**. I have always had a vivid imagination, which drew me to writing. Coming from a musical background, I love how poems allow words to flow like a melody. I am a sixth-year student at Maynooth Community College where I enjoy English as well as scientific subjects. My hope for the future is to study law or medicine allowing me to pursue my interests in English and science. I am also heavily involved in sports and eager to keep advancing my skills both in Irish dancing and athletics.

Dr Samuel Terwase Udogbo is a Catholic priest with the Holy Ghost Congregation (Spiritans). He is from North-Central Nigeria and has a keen interest in development studies. He works with international aid agencies on a variety of development projects in Africa. He is a social mobiliser and PhD candidate in Sociology at Maynooth University. He currently lives in Dublin, and hopes to work with a wide range of civic and political organisations in Africa after completing his studies.

Jessica Traynor is a poet, dramaturg and creative writing teacher. Her debut collection, *Liffey Swim* (Dedalus Press, 2014), was shortlisted for the Strong/Shine Award. Her second collection, *The Quick*, was a 2019 Irish Times poetry choice. She is Poet in Residence at the Yeats Society, Sligo, and a Creative Fellow of UCD.

Jay Vergara – I am 16 and I was born on the 16th of March. I am in fifth year in Maynooth Community College. My interests are music, art and writing. I can play the piano, which I first started at 7. I like to write poems to reflect how I feel sometimes. After college I would like to be a game developer.

Conor Walsh – I am a student in final year in Maynooth Community College. I am an avid piano player and also enjoy participating in the local productions with our drama group, The Teresa Brayton Players. I play for the local hurling team, Cappagh, and help to coach a basketball team with Kilcock Basketball Club. In the future I hope to enter the legal profession.

Glen Wilson is a multi-award-winning poet from Portadown. He won the Seamus Heaney Award for New Writing in 2017, the Jonathan Swift Creative Writing Award in 2018 and The Trim Poetry competition in 2019. His poetry collection *An Experience on the Tongue* is out now with Doire Press. Twitter @glenhswilson

Dr Owens Wiwa is an Executive Vice-President, Regional Director of West and Central Africa and the Country Director in Nigeria for the Clinton Health Access Initiative (CHAI). Dr. Wiwa heads CHAI's Nigeria office and plays a leadership role in health policy development and implementation at the Federal and State levels. From 1998 to 2007, Dr Wiwa worked at the University of Toronto, designing and leading research projects that focused on infectious diseases (especially HIV), and community and mental health in Africa and Nigeria. Prior to CHAI, he worked extensively as a physician in rural Nigeria and as a human and environmental rights activist with organizations such as Sierra Club and Amnesty International. He continues his struggles for justice as an Ogoni activist. Dr Wiwa has an MPH from Johns Hopkins University and an MB BCH from the University of Calabar.

Joseph Woods is a poet and writer and was for many years, director of Poetry Ireland. He has been living in Zimbabwe with his family for the past five years. His fourth poetry book, *Monsoon Diary*, was published by Dedalus Press in 2018.

PART 1

Essays

Remembering Dede and the fight for Justice

Why Ken Saro-Wiwa must be exonerated and remembered as a man of Peace

Owens Wiwa

I grew up in a small town called Bori in Ogoni. My father worked in the civil service and also managed the markets in Bori. My mother was a trader in the market. My father had many wives and I grew up in a household full of children. Our respect for our parents and older siblings was very important. I did not actually know my eldest brother Ken very well when I was growing up. He was away at school and I saw him during the holidays. There was always the noise of a radio coming from his room. He woke up very early and he put on the radio and listened to foreign language stations.

I studied medicine and worked in different places in Nigeria. After my internship I was posted back to Bori. One day Ken came to Bori to see our mother. He asked me to come with him in his car. He stopped by my clinic. We noticed a huge cloud of smoke bellowing out from the Shell plant. I told him that sometimes the smoke was so bad we weren't able to see a person 10 metres away. He asked why I didn't do anything about it.

"I treat those with asthma and bronchitis and other respiratory diseases due to smoke inhalation," I said.

"Do you think that's enough?" he asked.

He thought that I shouldn't just treat, I should prevent the pollution, which is basically what he was trying to do. That was my introduction to the whole world of government injustice. From then I started going to meetings and contributed in whatever way I could to the debate. Ken wrote the Ogoni Bill of Rights, which stated the rights of the Ogoni to have a voice within the Nigerian context. The Bill demanded the right to a clean environment and also demanded that Shell clean up our lands and pay compensation to the people whose livelihood was destroyed.

Ken was organizing peaceful protests. On the fourth of January 1993, over 300,000 Ogoni people marched from different villages and converged in Bori which is the political capital of Ogoni. It was a peaceful protest, asking the Nigerian government and Shell to read the Ogoni bill of rights and enter into negotiations with MOSOP. It was a simple request by a lot of people in a very peaceful way. While Shell claimed that after that they withdrew from Ogoni, I don't understand this. Shell's presence was in Ogoni in the form of their pipelines, drilling the oil from our land.

After the protest in January in Bori, in March or April the same year, some women and men went to their farms early in the morning. They met their farmland being bulldozed by contractors who were working for Shell and they protested immediately. Shell had the military, who were armed with automatic weapons come and they shot into the protesting crowd. A lot of people were injured, including one woman who lost an arm.

After the protest, Ken was subject to arrests, detention and several negative activities at the airports. He continued to voice his concerns about the destruction of the Ogoni environment. By polluting our land and waterways, Shell and the military dictatorship, were destroying the Ogoni people's livelihood, because we are farmers and fisherman. If you destroy our farms nothing is going to grow. If you pollute our rivers, the fish are not going to thrive and the fishermen will have nothing to catch. There will be nothing else to do, no jobs, nothing. After he started talking and writing about this, Ken was a marked man.

One day in August 1993, I was in my clinic as usual and motorcycles and cars were bringing in some people with gunshot wounds. I asked what was happening and was told the village of Kaa was invaded by armed men. I did what I normally do and attended to the injured. I then went off to Kaa to see what was happening. When I got to Kaa almost all the houses in that village had been either bombed out with grenades or burnt to the ground. There was a huge surge of people who were moving out of the village with almost nothing on them. Women, men and children were streaming out of the village to other villages nearby. While the national press tried to present this as an ethnic conflict, that was not true. There were also other incidents.

In May 1994, I was in my clinic and somebody ran in and told me that my brother Ken had been arrested by the Nigerian army. Four chiefs were having a meeting with other prominent people in Giokoo and they were murdered. My brother was in his house in Port Harcourt far far away from the murder. There are some people who believe the chiefs were killed by other chiefs in that community. There are others who say they were killed by the military in order to frame my brother and others. My brother was arrested in his house and several others were looked for and arrested. Journalists, writers, lawyers, doctors and teachers were specifically targeted. When I heard I went to search for him and some of my other friends. I didn't see them in the detention centres so I took a flight to Lagos because when he was arrested in 1993 he was brought to Lagos. I thought maybe the same thing had happened. The next morning, in Lagos, I heard on the radio that I was also wanted in connection with the murder of the chiefs.

The last time I saw my brother was two weeks before he was arrested on 22 May 1993. It was a normal meeting, but I remember vividly what he said went I saw him in his office in Port Harcourt; he said that he wanted me to stop being involved in the struggle. His words were, "those who we are dealing with are dangerous and they will stop at nothing". Ken thought they would want to kill me and that would hurt him and make him stop. He said it would be good for one of us to be alive. I was unhappy then because I was already so involved with this struggle. I came into this trouble a little bit reluctantly but the more I got involved, the more I had to stay involved. My life, and my outlook on life changed. Eventually, I was also one of the Ogoni professionals declared wanted for the murders that he was arrested for.

I went underground for eighteen months. I got quite a lot of help from a variety of people and organisations, including the Nigerian pro-democracy movement. Journalists in Nigeria helped. I was also in contact with those in the diplomatic community. Sister Majella was still in Lagos and she introduced me and my family to an American at the University of Lagos who hid my wife and small child in her house. I was also in contact with the International community by many ways, fax back then was the mode of communication and sometimes phone. During those eighteen months, I probably stayed in at least twenty-five different places.

During this time Ken was in detention and was able to write letters. He wrote letters to me to contact some lawyers which I did and his lawyers were not even allowed to enter the court sometimes. The trial was a sham. It was a military appointed tribunal with military orders to do what they were asked to do. His lawyers were not given a free hand to represent him so he asked them to stop. So they stopped and the government just went on with the charade.

My wife, at that time, Diana, was able to go to Ken in detention with my little son Befi and he would give her some letters. Some of the letters she would put in Befi's diapers and bring out. Some were put on plates of

Fishing livelihoods destroyed

food and the only thing that gave him some sanity while in detention was the fact that he could communicate and write. He wrote tons and tons of letters. Literally I was getting letters every week. So many letters to so many people. Letters were pouring in to what was his office in Lagos which I used to go to. Remember, I was underground and I would get these letters and give them to an activist who was not wanted and he would take those letters back to Port Harcourt. He would leave the letters for my wife or anyone who had access to Ken to give him his food and they would take them to him. Immediately he got the letters he would read all of them and respond. Some letters he kept and some were taken to his office in Port Harcourt. I think that one of my unfinished businesses – and I have quite a lot – is to actually go in search of these letters that people wrote to him. One of the people he wrote to was Sister Majella and those letters came to me and I passed them on to Sister Majella. That relationship was very important to him. You can see that through the letters.

I saw them together a couple of times and I was shut out from that communication. I am a very practical person. If somebody is ill, I ask what's the problem. I look for the signs and I treat. They were talking intellectually, theories and all those sort of things and it would go on and on, back and forth and very lively discussions. My brother was happiest when he saw her and they talked. I think it brought quite a lot of things to life for him. She was somebody that he could relate to intellectually and academically.

You can imagine during the struggle we were talking to everybody, we were with everybody and she was to him fresh air from a different context.

I was in Lagos when I heard my brother had been hanged. I thought how could God let this happen? Where was God? Up to that time (before I was underground), I never had the time to study what was happening in a global context and to see what sort of powers we were confronting. During Ken's time in detention and when I was underground, I read so many books and there was a book I read, that detailed the power of the Seven Sisters. When I read that book and saw what oil companies had been doing from Iran in the Middle East, toppling governments, doing all sorts of things, I knew my brother was in big trouble. I was drawn into this struggle by the fact that I wanted to treat people who were sick and to save lives. I saw the poverty when I was practicing. People were so poor they couldn't buy food to eat well and protect themselves against the most common illnesses; I just wanted that to change. Later, I got to read extensively about how much profit companies like Shell were making. I got to read about how much money the people in government were making through corrupt practices related to oil extraction and I said "Oh God this is not just about the cleanup." Quite a lot of toes had been stepped on. I thought – detain us, stop the press from publishing what we write, whatever, but don't kill us in order to stop the debate, which is what they tried to do.

After the executions it became obvious that we had to leave. Most of those who had been hiding us were frightened for their own lives for housing us. We didn't want to get them into trouble so I went to meet one of the diplomats that Sister Majella had introduced me to. I gave him my certificates and some of the small belongings that we had and he said he would get these to Ghana. We had to find our own way to Ghana so we literally crossed the Nigerian border on foot, went to Ghana, got our things and stayed in Ghana for about three weeks. I tried to get a Canadian visa, it wasn't the easiest in the world. However, Anita Roddick, the owner of the Body Shop, heard we were in Ghana and she helped us to get a British visa and got us on a plane from Ghana.

I joined the media campaign that was already on in London and amplified the volume a bit when I got there. They did save other lives. There was more spotlight on the other Ogonis that had been detained with my brother but were not hanged. A lot of other campaigns were also going on elsewhere, including the Ogoni Solidarity Ireland group and Sister Majella in Ireland. All of these campaigns contributed to their release from detention. It also helped to secure refugee status for Ogoni people who had to flee the detentions and the killings in Ogoni. They were accepted by many countries around the world to start a new life.

After Abacha died in 1998, there was a need for my parents who were still alive, and others who were psychologically and emotionally distraught with what had happened, to have a sense of closure. I had to look for the body of Ken, and to have a symbolic burial for him, for those left behind. We started the struggle to get his bones, his remains and the remains of the other eight who were hanged with him. We did not get the remains in time, so we had a huge symbolic burial where thousands of people turned up.

In 2000 Ken's body was exhumed from the grounds of the cemetery, a short distance from the prison where he was executed. I think that finding his body was such a relief to me and to all members of the family. We also found the bodies of the other eight hanged with him. This came about through help from lots of friends and the U.S. group Physicians for Human Rights. The exhumation and the DNA identification and the return of the bodies took four years and it wasn't easy. We wanted Ken's body back to Ogoni to give him a decent burial; to get closure for his children, for his wife, for our parents and for ourselves, that was number one. The second reason was that he loved Ogoni, and to have his remains outside of Ogoni I know would pain him.

After his burial I felt I had a sense of some fulfillment, some filling of a huge gap that had existed in my life since he was killed. I took a step back and decided to observe the struggle from afar and to study it more, to prepare to come back. Some have interpreted that to be a sort of silence and yes I have been silent for quite a while. So when I heard about a possible book and I saw the name, *Silence Would Be Treason*, I started asking myself, have I been silent for too long? I am on another journey right now where I'm saving lives, which is what I was taught to do in school.

I am very grateful that one part of my brother's life is going to be preserved by Maynooth University Library. This is something we are not very good at in Africa as a whole. When Sister Majella wrote to us, telling us what she wanted to do, the family were extremely grateful. I think this archive has some of the most important writings of Ken's moments in prison. It also has quite a lot of the material that was produced during the struggle, like the Ogoni flag. There is not much around now, even pictures.

MOSOP flag and cap

During the military crackdown in Ogoni many people hid Ken's picture and his books. They buried them in the ground and did not bury them in things that would preserve them, so many things have been lost. I am so grateful to the Maynooth University Library for preserving his letters and other things. I am very very grateful to Sister Majella for keeping his letters safe. Like I said there are so many letters to be looked for and to gather if they are still there. This is important. It brings back memories, some very good memories of that time and some difficult memories of that time, but to preserve it means his message survives and that is important.

The full interview, conducted by Dr Anne O'Brien, from which this article is excerpted, is freely available via the Ken Saro-Wiwa Audio Archive, which was created as a collaboration between by Maynooth University Library and Kairos Communications.

Noo Saro-Wiwa in Conversation with Anne O'Brien

On 10 November 2015, Noo Saro-Wiwa visited Maynooth University to launch the Ken Saro-Wiwa Postgraduate Award and to read from her book *Looking for Transwonderland: Travels in Nigeria.* I interviewed Noo for the Maynooth University Ken Saro-Wiwa Audio Archive, a collection of recordings of people connected to Ken Saro-Wiwa, including his brother Dr Owens Wiwa and Sister Majella McCarron (OLA). This piece is based on that interview.

Noo left Nigeria in 1978, when she was almost two years old. "Growing up in England it was all about watching cartoons, and getting on our bicycles and riding around the neighbourhood. Just very ordinary, everyday things, watching TV adverts, and seeing board games that we wanted – sometimes we'd get them for Christmas and that was amazing. Yeah, very ordinary and fun," she recounted.

While his family lived in England, Saro-Wiwa worked out of Port Harcourt.

"My father would come over every couple of months or so. He'd bring us chocolates, if he'd come via Switzerland he'd bring us these nice Swiss chocolates. You'd come back from school and he'd have his own homework set for us. He'd go out and buy textbooks and we'd have to sit down and work on these textbooks. Mainly to do with English language and things like that. He believed that idle hands are the Devil's hands. And so, sometimes he'd come into your room and just give you a book to read. I often really enjoyed the books, I discovered authors that way."

During the summer holidays Ken brought his family back to his home village, Bana in Ogoni. "I was never a massive fan of the village. Those visits highlighted the fact that we children weren't fluent in Khana. Obviously as an adult, going back, we saw areas of the village that were really beautiful. There was the river, which you'd approach from a height and you were looking down and it was fringed with mangroves. Really stunning, and it is such a shame that such a beautiful area, so full of wildlife and animals and plants, is neither a tourist destination nor a healthy agricultural area. That's a real tragedy."

Noo was aware of the dangers her father faced. "When I was 16, he wrote a letter to me when I was at boarding school, and he said that the military government could kill him. I was really angry because I thought he was being over-dramatic and scaremongering. But he knew the risks, and made sure that we knew the risks. But I guess I really didn't understand what he was up against until the actual night he was killed. He went to prison in 1993 and he was there for a month, and then he was let out again. Nigeria was that kind of place; it was a military dictatorship, people were in and out of prison. But also my father had a way of not making light of things, but you know he took things on the chin and he didn't allow being in prison stop him from focusing on what I would consider relatively trivial matters. In the archive here in Maynooth University, I was reading his letters to Sister Majella and he was talking to her about the Troubles in Northern Ireland and how heartened he was that some progress had been made in negotiations. That's what he was like, he didn't focus entirely on his predicament. He had an interest in what was going on around the world, and within the family. When someone is like that, they make it easy for you to underestimate the danger ahead."

Noo was 19 when her father and eight colleagues (the Ogoni Nine) were executed.

"I was a second-year university student at King's College London and he was sentenced to death – my father and his colleagues – on 31 October. So that came as a real shock, but then the international community really rallied round, so come 10 November it was just another day within that particular period. I must have attended classes, and then I went and did some shopping and I came back to the house where I was living in North London. My housemate had left a message, just a handwritten note on the table saying 'call your mother' and so I called my mother and she was the one who told me. I just put down the phone, which was the same reaction I had when I was told that my little brother died two years previously, I just put down the phone. And then went home immediately to my mother's house and spent the evening with the family. My cousins and my aunt and uncle came over."

Noo returned to Nigeria in 2000. The bodies of the Ogoni Nine had been dumped in unmarked graves.

"It was my father's 'mock funeral', as it were. I think in Ogoni culture people don't feel that a spirit's soul has rested until the person has been buried and so we thought we'd go through the motions of a burial even though we hadn't received his remains at that point. So we went back for maybe a week and a half. And then the next time we went was five years after that, in 2005, and by then we did have his remains and we buried him properly."

When she was 25 Noo decided she wanted to be a travel writer.

"I hadn't really been back to Nigeria apart from those two brief visits for my father's mock funeral and then the actual burial. Time had passed, and we now had a democracy. The military dictatorship was gone. There was an element of optimism within Nigeria, and it seemed a more hospitable place generally. Because so much time had passed, I felt ready to go back and I was really curious because I had spent a lot of time travelling in other countries around Africa on holidays as well as writing travel guidebooks. It suddenly occurred to me that actually Nigeria is a place that I could travel around in the same way that I had done in Ghana, and Madagascar etc. That realisation was actually quite exciting. The idea that I could just jump on a minibus and travel from city to city was a really novel idea. I thought this was a really good way of re-connecting with Nigeria, and doing it that way didn't remind me of my father's death. Riding on camels in Kano, getting on boats along the river outside Calabar, being in the mountains in Obudu; doing all those sorts of things, I saw a different side to Nigeria and I was really able to disassociate being in Nigeria with my father's murder. So it was a great trip in that respect. It was infuriating

Noo Saro-Wiwa and Ciara Joyce

in its own way, but I really really enjoyed it. It took the sting out of the word 'Nigeria' for me. It was a sort of therapy that I needed."

Noo visited the Maynooth University Library exhibition to mark the 20th anniversary of her father's execution and viewed a number of items – including her father's letters.

"I was just really amazed and grateful that there are people here dedicated enough to put together an archive like that. You know, just preserve my father's memory in that kind of way, it's wonderful. It really reminds you of his struggle. Even though you're aware of it, you think about it every hour of every day, seeing his writings there and having the audio recordings, of my uncle Owens and Sister Majella, you know it really brings home to you what the struggle is all about. It's wonderful, I'm so glad we have this resource. I really like the tablet screens in the exhibition area where you can look at his letters and some of the photographs in Ogoniland that Sr. Majella McCarron and her colleagues took at that time and other items. So you know it's a wonderful way of capturing that part of history and reminding people about the struggle. I'm so so grateful and I want to spread the word about this archive."

The full interview, conducted by Dr Anne O'Brien, from which this article is excerpted, is freely available via the Ken Saro-Wiwa Audio Archive, which was created as a collaboration between Maynooth University Library and Kairos Communications.

'...and don't ever forget the Ogoni People'[1]

Majella McCarron

Introduction

I returned to Ireland from Nigeria in 1994. I had spent 30 years as an educator in a newly independent country (1960), with eager students seeking the fulfilment of their many and varied capacities.

I'm from Fermanagh and in my time in Nigeria, I was always concerned about the border conflict in Ireland. I had hoped that my sabbatical year could be devoted to assisting with its resolution. However, I felt morally compelled to try the save the Ogoni Nine from execution. I was unsuccessful and they were hanged on 10 November 1995.

Colonial Constructs

Nigeria is a colonial construct. It was a British colony from 1901 until independence in 1960, making it 60 years old in 2020. Such constructs are intended to be of benefit to those European rulers who drew the lines on a map sitting around a table at the Berlin Conference 1884-1885[2]. Africa today is a pattern of artificially created modern states, created by force or barter. There is another map upon which Africa is imposed: a cluster of indigenous populations with discrete geographies, climates, languages and cultures. The Unrepresented People and Nations (UNPO) was founded in The Hague in 1991 as a platform for such political voices. The Ogoni is one such indigenous population lying on the Atlantic seaboard. It rests on a delta rich in soil that feeds crops and fish in abundance. Far underneath this delta are the oilfields. The exploitation of these oilfields exudes much that is harmful to life in various forms – gas flares, pipeline leaks, the loss of animal life and the destruction of farming on land and sea. At the same time the exploitation of the gas fields makes millions of dollars for Western oil multinationals. Alongside the destruction of livelihoods is the related destruction of cultural norms and traditional lifestyles: and all this in 60 years!

Saro-Wiwa served as vice chair of UNPO for a time. He marked the opening of the UN General Assembly's International Year of the World's Indigenous People in 1993 on Ogoni Day, the fourth of January. By the

1 Maynooth University Ken Saro-Wiwa Archive, Letter dated 24 July 1994 MU PP/7/5
2 The conference contributed to heightened colonial activity in Africa by European powers.

same date the following year, 1994, he was under house arrest and entering what was to be his final year.

Natural Resources

Ken Saro-Wiwa was born into this colonial construct in 1941. His family were busy people in the traditional sense. They endured the oil world around them. As a schoolboy Ken slowly became aware of the destruction of his small homeland. The oil companies had arrived when he was ten. The gas flares began to fill his childhood sky. Crops began to wilt. Oil was extracted from the ground and piped away by men and machines he saw only briefly. Ken became a writer: these realities began to invade his poetry and prose, his plays, his political commentaries. There was a keen sense of sadness and loss permeating the humour. He loved the Ogoni people and died for their wellbeing. Those who took his life ignored the powerless reality under the colonial construct. They were frightened: his name could not be spoken nor could people gather to mourn on the day he was hanged. That is why I wrote the poem *10 – 11 – 1995 A Night of Death*, which is included in this volume. It is why we are remembering that same day and night 25 years later. We stand as witnesses to colonial pain as destruction reigns. Shell 'vultures' still stand. Greed and profit disrupt the human population.

"Milking" the Constructs

The suppression of the Ogoni in 1995 was vicious. It began in 1990 where in anything but an aggressive mood, Ken Saro-Wiwa took an Ogoni Bill of Rights[3] for discussion with the military ruler Sani Abacha of the day. He then went out with his people on a peaceful protest march and later prayed at an all-night vigil. Did he ever imagine the ferocity of the reaction from the military dictatorship?

The ferocity of the military dictatorship was matched by the might of Royal Dutch Shell calling on the dictatorship to push oil pipelines through small fields cropped by women's manual labour in the heat of the sun. One young man was shot in the back and died, while a woman had her arm blown off by accompanying soldiers of the specially trained force under another fierce commander, the Rivers State Internal Security Taskforce.

An extraordinary murder of four Ogoni people, important in their own right, with two related to Saro-Wiwa, took place during the day of

3 Ogoni Bill of Rights: http://www.bebor.org/wp-content/uploads/2012/09/Ogoni-Bill-of-Rights.pdf

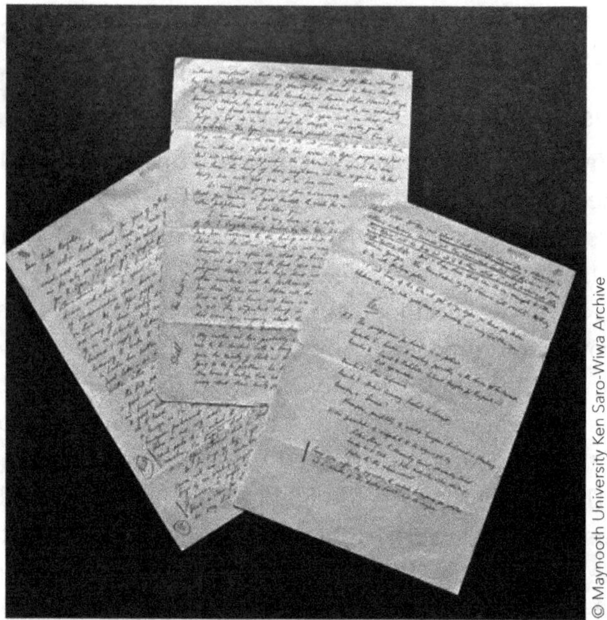

© Maynooth University Ken Saro-Wiwa Archive

Letters

21 May 1993. Ken Saro-Wiwa was arrested, accused of the murders and, with others, held until he was hanged, with eight others, on 10 November 1995.

It was suspected that the destruction of ten or more Ogoni villages was the work of trained forces under the guise of enemies from neighbouring clans. Slowly the destruction petered out. A peace deal was signed as a way of closure rather than of justice. The Ogoni people knew escape to freedom or a better life was now a priority. Shell wanted to resume operations. That did not happen. It has been keeping on the pressure for the last 25 years. Shell offered to pay the bill for the United Nations Environment Programme (UNEP)[4] report despite it being a joint partner with the Federal Government. That was in 2011, 16 years after the executions. There was relief as the right to clean water was highlighted as paramount in the UNEP report and the removal of cancer inducing contaminants seemed imminent. It was such a relief for me and taught me to trust the people where truth is life. A clean-up is ongoing, but there is much confusion about its quality and fairness. Nine years have passed since the dangers of pollution were universally acknowledged.

4 UNEP (2011). Environmental Assessment of Ogoniland report: https://www.unenvironment.org/explore-topics/disasters-conflicts/where-we-work/nigeria/environmental-assessment-ogoniland-report

Resisting Resource Exploitation

In his essay in this volume, Dr Owens Wiwa describes the events around his brother's arrest and his life as a fugitive. Owens contacted Shell Lagos to plead for his brother's life. The conditions Shell wanted, led him to turn down their offer, as he knew his brother would have done. I had come to know Owens on visits to help with the relief effort in the villages in Ogoni. I had facilitated assistance from the EU, Trócaire, the Diocese of Port Harcourt and the Daughters of Charity. The first letter in the archive is a letter of thanks I received from Ken[5]. He advised me to meet his brother in Ogoni while he stayed behind at his office in Lagos. Later on my return to Lagos, Ken took a picnic container filled with vaccines that I had sourced from the United Nations High Commissioner for Refugees (UNHCR), on an early morning flight, to Owens in the Ogoni villages. I felt it was all so futile confronted by the enormity of punishing power and tried to capture my feelings in my poem *Dying Village*, which is included in this volume. Everyone around us in Lagos and Port Harcourt was terrified. Meanwhile, Ken was writing letters and those I received were donated to Maynooth University Library in 2011 and published as *Silence Would be Treason: Last Writings of Ken Saro-Wiwa*[6].

I returned to Ireland in August 1994. It seemed cruel to leave but no one really wanted me around. It was believed the military were close by. I sent Ken a number of letters from Ireland while he was in detention. Many people have asked if my letters to him have been kept somewhere.

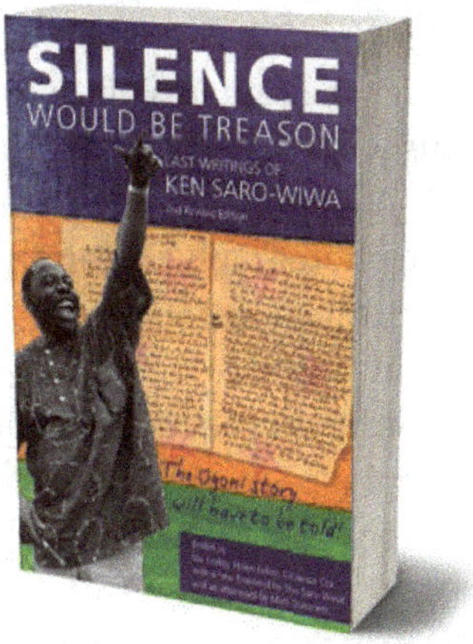

5 Maynooth University Ken Saro-Wiwa Archive, Letter of 20 October 1993. PP/7/1

6 Corley, Íde, Fallon, Helen, Cox. Lawrence (2018). *Silence Would be Treason: Last Writings of Ken Saro-Wiwa.* Senegal: Daraja Press,. http://mural.maynoothuniversity.ie/10161/

I did not know until a new book *The Politics of Bones*[7] suggested the answer ten years later: Saro-Wiwa had a standing instruction about filing all documents sent out from his detention. This was not always obeyed and much material was seized and destroyed during raids on his office; it is presumed my letters were among the items destroyed.

After the executions, many Ogoni people, including Owens Wiwa and his family, left on the paths and in the buses from Lagos to Cotonou in the next door Republic of Benin, then on to Togo and Ghana. Benin set up a United Nations High Commission on Refugees (UNHCR) camp which was to house about 1,000 Ogonis. There are photographs from the refugee camp in the archives at Maynooth University.

Twenty-five years later there are some Ogoni refugees still reluctant to risk returning home. I kept in touch with some of them including catechist Anthony Koteh-Witah. He was to go on to the US where he is now a Capuchin friar in Detroit. Owens Wiwa and his family went from Accra to London, where I joined them from Dublin for a few days. Such was the flurry of activity, that there was not even a breather to restore their baby's shoe, dropped along the way. Home was not forgotten with the formation of MOSOP USA, MOSOP Canada, MOSOP UK and possibly more, active in 2020.

Rest in Peace

In 2002 Ken's parents received me well and took me to visit the grave site waiting the release of Ken's body for burial. The account of Owens's efforts to retrieve his brother's body, in *The Politics of Bones*[8], is very moving. It took almost seven years to recover the remains of the Ogoni 9, who had been buried in unmarked plots. This work was aided immeasurably by the group Physicians for Human Rights[9]. A signet ring with one of the remains was identified from a large portrait photograph of Ken in his sitting room. Nobody could remember such a ring, much less the watch with the blue plastic strap. The watch was inscribed with the words "Human Rights, Vienna 1993."

At a UN meeting which he attended, Ken had breakfast with Mairead Corrigan of the Peace People.[10] She, along with others, secured a Nobel Peace nomination for him in 1995. His son Ken Junior (who died in 2017,

7 Hunt, Timothy (2005). *The Politics of Bones: Dr Owens Wiwa and the Struggle for Nigeria's Oil*. Toronto: McCelland and Stewart.

8 Hunt, *op.cit.*

9 Physicians for Human Rights investigate and document human rights violations: https://phr.org/

10 The Peace People began in 1976 as a protest movement against the ongoing violence in Northern Ireland. http://www.peacepeople.com/

aged 47) received the nomination from Corrigan in Belfast, one month before his father's death.

The United Nations Environmental Programme (UNEP) Ogoniland Report was the burst of truth which verified all we had heard. That happened in 2011 and, parallel to Shell's stubbornness, is nine years struggling to assist the people, 25 years after Saro-Wiwa's death. It verified the worst instances of oil pollution in the observations and testing carried out by world class scientists. The administration of the pot of money allocated for the clean-up is causing ongoing friction.

Finally, in this essay, I would like to pay tribute to Ogoni Patrick Naagbanton, who died in a road accident in Port Harcourt in 2019. His tribute to *Silence would be Treason* reads:

"This book contains information about the heydays of the Ogoni struggle, its victories, failures, betrayals and travails in the naked face of highly organised state/corporate violence and conspiracies against a marginalised and embittered people of the eastern Niger belt in Nigeria.[11]"

Naagbanton, founder and director of the Centre for Environment, Human Rights and Development (CEHRD)[12], was the guest of Frontline Defenders[13] in Dublin in 2010. Writing about his visit in a local newspaper on his return to Nigeria, he mentioned his mother's appreciation of his own successful delivery and care by Irish midwives.

May Patrick Naagbanton rest in peace.

More information on Sister Majella McCarron's early life and her work on justice and peace can be found in a series of eight audio recordings she made for the Maynooth University Ken Saro-Wiwa Audio Archive.[14]

11 Corley, *op.cit.*

12 CEHRD (Centre for Environment, Human Rights and Development was founded to respond to the environmental, human rights, rural health, and underdevelopment problems plaguing the Niger Delta. https://cehrd.org.ng/

13 Front Line Defenders or The International Foundation for the Protection of Human Rights Defenders is an Irish-based human rights organisation founded in Dublin, Ireland in 2001 to protect those who work non-violently to uphold the human rights of others as outlined in the Universal Declaration of Human Rights. https://www.frontlinedefenders.org/

14 Maynooth University Ken Saro-Wiwa Audio Archive: https://www.maynoothuniversity.ie/library/collections/ken-saro-wiwa-audio-archive

Global solidarity: The translocal connection

South to North Ogoni to Erris

Majella McCarron

Introduction

In 2000, at the beginning of a new millennium, and five years after the execution of the Ogoni Nine, Royal Dutch Shell was about to propose its planning application to the Irish Government for an offshore oil well, eighty miles out to sea off the coast of Mayo, in the west of Ireland. A pipeline from the oil well would transport gas to the mainland with the final stretch passing through local villages. Many people objected and suggested alternative routes. Protests resulted in much mental and physical suffering. Court cases were fought, people were sent to prison and dozens brought before the Courts. Finally, the pipe was encased in a tunnel and laid below the seabed at the minimum distance advised for safety in case of a blowout.

Ogoni Solidarity Ireland

With my experience of the Royal Dutch Shell approach in Nigeria and then working in Dublin on such issues, I set out to investigate if the approach used differed from south to north, Ogoni to Erris. My interest was academic until I began to observe and learn and finally, to report and speak. I began to share my experience casually at first and then as a member of Ogoni Solidarity Ireland.

Ogoni was largely unknown in Ireland when I had approached the *Irish Times* in January 1995, four months after returning from Nigeria to Ireland, to publicise my pleas on behalf of the Ogoni people.[1] On confronting Shell the local Ogoni farm men and women were threatened with 'wasting' operations, displacement, imprisonment... as the price for minimal non-violent protest by a special military unit organised for this purpose. The following month, February 1995, following the *Irish Times* article, there was discussion about what was going on in Ogoni, at the annual Afri (Action

1 MacConnell, Sean (1995). Nun claims proof of Nigerian plot to suppress ethnic group. *Irish Times*, 16 January, p.4

from Ireland) human Rights Conference in Kildare: this led to the establishment of Ogoni Solidarity Ireland.

The Bogoni

In 2001, Ogoni Solidarity Ireland suggested holding its fourth Annual Ken Saro-Wiwa Memorial Seminar in Geesala on the Erris peninsula in Mayo. Maura Harrington, Komene Famaa, an Ogoni refugee, MOSOP member and Irish scholarship holder, and myself put a programme together for June 2001. One of our guests was Mrs Joy Phido from MOSOP UK. Maura recalls the event:

> The seminar was titled 'Corrib Gas – Great Gas for Whom? A Multinational Impact Assessment'. This seminar provided the first public information forum in relation to the Corrib Gas Project which was not controlled and managed by the oil companies, who refused an invitation to attend. The late Richard Douthwaite made a prescient presentation to the seminar, as pertinent 20 years on as it was then. Many who attended that seminar subsequently worked together as part of the Shell to Sea campaign. Komene Famaa and Joy Phido were the first Ogoni people met by Erris people – an instant empathic bond was formed which has strengthened over the years.

Both the people of Erris and the people of Ogoni live close to the ocean on special habitats of bog and delta. This was recognised by the people of Erris who began to address themselves for a time as the Bogoni. They erected large colourful Ogoni murals, one such (12 by 8 feet) was locally constructed by Terence Conway and painted by John Monaghan and Chris Philbin.

The following year, 2002, Owens Wiwa travelled to Mayo. En route from Shannon, Dr Wiwa met with the then Labour TD, now President Michael D Higgins, and urged that the State should not 'negotiate away the environment'. Unfortunately Dr Wiwa's warning comments have proved true:

> '... then there is the social tension that can arise between those who do and those who don't want this sort of economic development – the splits it can cause in families, the lack of transparency among the developers and politicians. I have seen all this happen before.'

In 2005, Owens Wiwa continued to show his support by being present to welcome the Rossport Five out of prison, having served a sentence of

Owens Wiwa and Majella McCarron

94 days for refusing to allow the company on to their land. Owens was one of the leaders of the annual Afri Famine Walk in 2006. His nephew Barika Idamkue had come to Dublin and Belfast to plead for Ken Saro-Wiwa's life in March 1995. The empathetic bond has indeed been strong. The homes of Erris, Dublin, Belfast and Fermanagh have been welcoming hosts to visiting Ogoni people.

Translocal Connections

While Shell to Sea and the Movement for the Survival of the Ogoni People built their mutually supportive links over 12 years, their respective websites were active before this and continue to act as vital resources. Shell to Sea has become a mentor for other groups in Ireland. The group Love Leitrim, when confronted with fracking in their local area, went to Erris for information. Leitrim is also in the west of Ireland, far from the capital, Dublin.

Love Leitrim was fortunate to have a researcher on site for most of the six years of its campaign. Jamie Gorman was very unobtrusive as a participant observer for a local case study. He completed his doctoral thesis in 2019 at Maynooth University, and it became a welcome source for the term 'translocal connections'.[2] He identifies it as one of four pillars of a successful campaign under the heading 'Connecting Frontline Communities'.

2 Gorman, J. (2019) Jumping scales and influencing outcomes: A case study of community development for environmental justice. PhD thesis, Maynooth University.

Visits to learn

Gorman records the decision of the Love Leitrim group to visit the Erris group and be informed of its experiences with multinational natural resource activity. One of his respondents describes the visit:

> '...seeing all the security and they all had their eyes covered and their mouths covered and bandanas and stuff...surrounding the car and taking all our photographs and writing down the number... "my God, this is our future!" ...'
>
> "We realise now at this stage that what we're doing has happened before and it's happening all over the world... the fact that companies are going into small communities and take their natural resources... so we learned all that."

Visits to support

Members of the Love Leitrim group felt impelled to travel the very short distance to support a neighbouring campaign because the same company was about to start preliminary moves towards fracking. Gorman refers to the necessary delicacy required in approaching another campaign even in a support capacity. He quotes a Love Leitrim member saying 'you don't want people to feel like you're stepping on their toes... they're the local people and they're doing things how they want but you're going to support them.'

The Love Leitrim group took turns on visits of solidarity to the Fermanagh site:

> 'just offering help or bringing up a bit of food if we could. That kind of thing, just working in the background...we went up with a car load of wood one time for the fires, kinda thing, just small little things in the background.'

Visits to inform

The research remarks on the local response to visitors and indeed gives a list of these. It points out those considered to have contributed greatly. International speakers were welcome, especially those who had first-hand experience of company reaction to campaigning, adding to local experience and analysis. One interviewee remarked that 'once you get people just spending time together it makes a huge difference you know? In the same room, in the same place for a while rather than just email.' Gorman suggests that such "translocal connections help frontline communities to amplify each other's struggles and demands." This is a very important conclusion about the value of solidarity.

Conclusion

This research affirms the thrust of human solidarity in the global context. It affirms my last physical act of solidarity on Ogoni Day, 4 January 1994. An international journalist based in Lagos and another local justice advocate were insistent that we travel on the morning flight from Lagos to Port Harcourt and on by road to Bori for the celebration of Ogoni Day. I was hesitant as I had heard the leaders of MOSOP had been placed under house arrest. This was true and on arriving we went to the office of the late Claude Ake, a close associate of Ken Saro-Wiwa, a Rivers State man and renowned academic in the social sciences. Saro-Wiwa was under house arrest. I quietly asked if I could be taken to him. The gate was firmly padlocked and there was a security detail around. Ken was called by a young relative who recognised me. We spoke for a few minutes from either side of the gate with Ken asking me to go on to Bori with my companions to tell the people that I had seen him - solidarity in the face of powerlessness.

I hope my correspondence from Ken Saro-Wiwa and other items I have donated to Maynooth University Library will be a source of inspiration for scholars and activists well into the future.

Legal redress for Ogoni Communities: how can oil majors be held to account?

by Daniel Leader

In his final speech to the military tribunal during his trial in 1995, Ken Saro-Wiwa spoke these words:

"I repeat that we all stand before history. I and my colleagues are not the only ones on trial. Shell is here on trial and it is as well that it is represented by counsel said to be holding a watching brief. The Company has, indeed, ducked this particular trial, but its day will surely come and the lessons learnt here may prove useful to it for there is no doubt in my mind that the ecological war that the Company has waged in the Delta will be called to question sooner than later and the crimes of that war be duly punished. The crime of the Company's dirty wars against the Ogoni people will also be punished."

Twenty-five years after Saro-Wiwa spoke those words, Shell is being pursued in the courts in England and the Netherlands by victims of its systemic pollution of the Niger Delta. This chapter seeks to explain how a new generation of activists and lawyers are seeking to hold Shell, and other multinationals, to account for the "ecological war" they have waged in the Delta.

Few situations exemplify the current challenge of corporate impunity as starkly as the widespread and systematic pollution of the Niger Delta. Most of the pollution emanates from the defective oil pipelines and infrastructure operated by subsidiaries of western domiciled parent companies such as Shell, Total, Exxon Mobil and ENI. The situation has worsened markedly since Ken Saro-Wiwa's death and the facts are stark: in 2006 a team of international experts estimated that between 9 million and 13 million barrels of oil (1.5 million tons) had spilt into the Niger Delta over the past 50 years.[1]

Overall, it is estimated that the inhabitants of the Niger Delta have experienced oil spills on a par with the 1989 Exxon Valdez disaster in Alaska every year for the past 50 years, a yearly average of about 240,000 barrels.[2] These statistics mask a human tragedy on an extraordinary scale. The pollution is ingested by local communities and seriously impacts human health

1 Niger Delta Natural Resource Damage Assessment and Restoration Project. Federal Ministry of Environment, Abuja. 31 May 2006.

2 Petroleum, Pollution and Poverty in the Niger Delta, Amnesty International [2009], p.16

and mortality rates. For example, a recent epidemiological study by the University of St Gallen in Switzerland found that infants in the Niger Delta are twice as likely to die in their first month of life if their mother's lived near an oil spill. That amounts to a scandalous 11,000 premature deaths per year.[3]

The causes of this systematic pollution are complex but include a lack of investment in pipeline maintenance and the necessary technology to prevent third party "bunkering" by criminal gangs who siphon off oil. Whatever the cause of the pollution, oil companies are required by Nigerian law to ensure that any oil spill is cleaned-up within 30 days.[4] This basic legal requirement is, however, largely ignored and unenforced.

Ogoniland has been the subject of particularly detailed environmental studies. In 2011, the United Nations Environment Programme's (UNEP) published a comprehensive Environmental Assessment of Ogoniland based on two years of field work during which 4,000 samples were collected from over 200 sites.[5] It reported that the Ogoni people were exposed to severe oil contamination on a daily basis, which impacts their water sources, air quality and farmland.

In a number of locations public health was significantly at risk. Some water wells tested by experts showed 1,000 times higher concentrations of oil contamination than permitted under Nigerian drinking water standards.[6] UNEP urgently recommended "the largest terrestrial clean-up operation in history," but ten years on the communities are still waiting.[7]

Why is this human tragedy allowed to persist while western oil majors continue to profit so richly from the oil which flows from under the feet of these devastated communities? The short answer is that the weak regulatory regime gives space to avaricious corporations who fail to maintain and invest in their infrastructure or clean up their oil spills. Few people pay any attention to the plight of the Niger Delta and while multinationals come under periodic pressure to clean-up their act, such pressure has not been sustained and has been of insufficient strength to affect real change.

3 *Effect of oil spills on infant mortality in Nigeria*, Proc Natl Acad Sci USA 2019 Mar 19; 166(12); 5467-5471
4 Sections VIII B.4.1 and VIII B.2.11.1 of the Environmental Guidelines and Standards for the Petroleum Industry in Nigeria (2002).
5 Environmental Assessment of Ogoniland (United National Environment Programme, 2011), http://postconflict.unep.ch/publications/OEA/UNEP_OEA.pdf
6 Ibid 11.
7 See for example 'No Progress: An evaluation of the Implementation UNEP's Environmental Assessment of Ogoniland' (Amnesty International, 2014).

Access to Justice in Nigeria

The communities affected by decades of systematic pollution have been provided with little assistance by the Nigerian courts or regulators. The difficulties in accessing justice in oil spill litigation through the Nigerian courts have been well documented.[8] There are multiple barriers to justice. In the first instance, the overall inefficiency of the litigation process can result in decades of delays as cases work through the archaic judicial system. Oil companies invariably use the system to grind down claimants and tend to appeal every adverse decision all the way to the Nigerian Supreme Court. The subsequent delays in the determination of cases are dramatic.[9] For example, in the case of *Tiebo and Ors v. Shell Petroleum Development Company Ltd (SPDC)*[10], a community sought damages arising out of an oil spill in 1988 and it was 17 years before the Supreme Court handed down its final judgment. Similarly, another oil spill claim, *Agbara v. SPDC*[11] was issued in 2001 and judgment was not handed down by the High Court until 2010 and it took a further 9 years for the case to go through the appeal courts – 19 years in total.

Equally, the mismatch in resources and power between oil companies and the communities they pollute renders any legal claim a near impossibility. Oil spill litigation requires specialist technical expert evidence to respond to complex evidential disputes, which is unaffordable for impoverished communities. In addition, claims are brought as representative actions by community chiefs so that, even in the miniscule proportion of cases which result in some level of compensation after legal battles which span decades, the damages are usually distributed to community leaders rather than to the inhabitants of those communities whose livelihoods have been destroyed by polluted waterways and farmland.

Finally, there are no reported cases in which a Nigerian court has actually ordered the clean-up of an oil spill by an oil company. There is one example of injunctive relief being granted to a community in 2005 with regard to gas flaring, a widespread and controversial practice in the Niger Delta. However, the oil companies flatly ignored the injunction and the

8 See, for example, Jedrzej Frynas, *Oil in Nigeria: Conflict and Litigation between Oil Companies and Village Communities* (1999)

9 In recent litigation in the English High Court concerning the enforcement of arbitration awards the Court of Appeal observed that the extreme delay *"result[ed] from the workings of the Nigerian legal system"* and has *"gone beyond the 'catastrophic' description adopted by Tomlinson J in 2008". IPCO (Nigeria) Limited v NNPC* [2015] EWCA Civ 114 at 166.

10 *Tiebo and ors v. SPDC* [2005] 9 NWLR [Pt.931].

11 *Agbara v. SPDC* [2010] Case No: FHC/ASB/CS/231/2001.

judge was ultimately transferred to another state.[12] The position is no better with the regulator. The Director General of National Oil Spill Detention and Response Agency (NOSDRA), the official regulator of the Nigerian oil industry, observed in 2012 that it is simply unable to enforce compliance because it has no power to sanction polluters.[13] As a result, given the hopeless prospect of accessing justice in Nigeria, impacted communities are increasingly seeking redress in the national jurisdiction, in which the parent companies of the Nigerian subsidiaries are domiciled.

The case of the Bodo Community

The Bodo Community in Rivers State was one of the case studies for the UNEP report and serves as an important example of the power of international litigation in holding oil majors to account. Bodo is a coastal community of about 30,000 people, which is situated in the heart of Ogoniland. Its predominant economic activity has been fishing since it borders 9,000 hectares of mangrove creek. In late 2008 two large oil spills occurred on the Trans-Niger Pipeline. The evidence showed that neither oil

Bodo Creek

12 *Gbemre v Shell Petroleum Development Co.* Suit No FHC/CS/B/153/2005, 25 July 2005. See also H. Osofsky 'Climate change and environmental justice: reflections on litigation over oil extraction and rights violations in Nigeria' (2010) 1 *Journal of Human Rights and the Environment* 189-210 at 191.

13 Energy Mix Report, 12 November 2013. http://energymixreport.com/laws-against-oil-spillage-weak-says-nosdra/

© Leigh Day

Bodo Creek

spill was contained or clamped for over 5 weeks on each occasion.[14] As a result, the community's experts contended that the spills, which destroyed over 2,500 hectares of creek, caused the largest loss of mangrove habitat in the history of oil spills and widespread economic and environmental devastation.[15]

In April 2011 the Bodo community withdrew their Nigerian claim, which had not progressed since 2009 and instructed Leigh Day as their English lawyers to issue the claim in the London High Court. The claim was brought against both Royal Dutch Shell plc (RDS) (the parent company domiciled in the England and the Netherlands) and Shell Petroleum Development Company Ltd (Shell Nigeria). The Claimants contention was that the RDS exercised de facto supervision and control over their subsidiary and that, therefore, the failure of Shell Nigeria to adequately maintain their pipelines, to contain the oil and clean up and remediate the pollution was also a failure of supervision and control for which RDS was directly liable. The claims were twofold: i) a community claim brought as a representative action[16] claiming clean up community land and compensation for the damage caused and ii) 15,601 individual claims for loss of earnings and nuisance by individual inhabitants of Bodo.[17]

14 Particulars of Claim, para 21.
15 http://www.leighday.co.uk/International-and-group-claims/Nigeria/Background-to-the-Bodo-claim
16 Brought pursuant to Civil Procedure Rules 1998, Part 19.6
17 Brought as a Group Litigation Order pursuant to Civil Procedure Rules 1998, Part 19.11

The claims were brought under the relevant Nigerian statutory provisions and the common law torts of negligence, nuisance and the rule in Rylands v Fletcher.[18] Shell Nigeria agreed to submit to the jurisdiction of the English courts and conceded liability with respect to the two oil spills that were the result of equipment failure. However, there was a significant factual dispute between the parties with regard to the volume of oil that had resulted from the spills. Shell contended that a total of 4,000 barrels had spilt but the claimants' experts put the figure at 500,000 barrels.[19] As a result, there was a fundamental dispute as to the extent to which the widespread damage to the creek could be attributed to the 2008 oil spills, as opposed to other oil spills that were caused by third party bunkering. The litigation, therefore, required the instruction of specialist technical experts to resolve the issue. Both parties instructed experts in satellite imagery, which ultimately demonstrated that the devastation to the Bodo creek had indeed been caused by the 2008 spills.

Shell had initially offered to settle the claim for $4,000, behind the backs of their Nigerian lawyers. In January 2015, because of the expert evidence that had been obtained to rebut Shell's case, the claims settled for £55 million. The damages were paid directly to each of the 15,000 claimants and transferred to Nigerian bank accounts that were opened in their names. This represented the largest damages settlement in the context of Nigerian oil spill litigation and has enabled the Bodo Community to rebuild economically. However, the fight for clean-up continues to this day. Since the 2015 settlement, the Bodo community has had to return to court twice, to force Shell to commence clean-up, and by 2020 only about one third of the creek had been cleaned up to international standards. There is a reason why the Bodo Community is the unique focus of an internationally led cleanup process by Shell – it is only the pressure of the international litigation which has pushed Shell to clean up at all.

Although, for reasons which are obscure, Shell Nigeria submitted to the UK courts in the Bodo case, they are now robustly fighting all international litigation. Further cases are being brought in the Netherlands *(Akpan v Royal Dutch Shell)* and by two new communities in the UK Courts *(Okpabi v Royal Dutch Shell)*, all of which is being fought tooth and nail by Shell who refuse

18 A copy of the Particulars of Claims can be found at http://platformlondon.org/wp-content/uploads/2012/06/The-Bodo-Community-and-The-Shell-Petroleum-Development-Company-of-Nigeria-Ltd.pdf

19 http://www.leighday.co.uk/International-and-group-claims/Nigeria/Background-to-the-Bodo-claim

to accept that the parent company carries any legal responsibility for the environmental negligence of its Nigerian subsidiary. The UK Supreme Court is due to deliver its judgment on this point in 2020 and, if successful, the impacted Nigerian communities may finally have the legal tools available to them to hold Shell and other multinationals to account for the first time.

Such international litigation is not a panacea, but in circumstances where multinationals have operated with complete impunity for decades, it is surely right that communities which have been the victim of environmental devastation should hold to account those companies who profit so richly from their oil. That is what Ken Saro-Wiwa foresaw in his final speech and, increasingly, it seems that he his words will have proven to be prophetic.

Ken Saro-Wiwa: the Ogoni youth hero

Samuel Udogbo

Introduction

We are living in a time of worldwide struggle. In many nations of the world, especially in Africa, the upsurge in social movement activity is most clearly marked by expressions and quest for freedom, self-determination, self-actualization, ethnic nationalism and resource autonomy and control. Nigeria for example, since independence, has experienced a series of ethnic struggles of various dimensions and magnitudes characterised with violent demonstrations, protest marches, civil insurrections, militancy and armed struggles. The Niger Delta in southeast Nigeria has been one of the hotbeds and most common examples cited in scholarly literature. The region has been targeted for crude oil extraction since the 1950s. There has been extreme environmental damage from decades of indiscriminate petroleum waste dumping – a major issue that has often led to peaceful and violent struggles in the region.

This essay draws on my Maynooth University doctoral research project which explored Ogoni's resistance in Nigeria. Based on my seven months ethnographic fieldwork and the Movement Relevant theory[1] adopted, I established a detailed analysis of how political marginalisation and repression by the Nigerian state and Shell Petroleum Development Corporation (SPDC) is the major factor for Ogoni's struggle for self-identity, environmental justice, resource autonomy and self-determination. There is enough evidence suggesting that the Ogoni struggle has shaken the Nigerian state and has also caught the attention of the international community (Kukah 2011; Okonta 2008), as was the intention of Ken Saro-Wiwa. For this essay, however, I concentrate on my experience with Ogoni youth, who, posthumously, consider Ken Saro-Wiwa as the Ogoni Hero: the man who spent his

1 The Movement-Relevant Theory: MRT puts the needs of social movements at its heart. Rather than reading the dominant social movement theory, this generates theory largely outside of academic circles. The MRT focus specifically on movement-relevant social movement theory, not social movement histories. This approach does not categorically reject earlier theoretical perspectives, but seeks to glean what is most useful for movements from these earlier works. Likewise, this emergent direction entails a dynamic engagement with the research and theorising already being done by movement participants. It bridges the divide between social movement scholarship and the movements themselves. It is social movement theory that seeks to provide 'useable knowledge for those seeking social change' (Flacks 2004, p. 138). for further discussion see Bevington & Dixon 2005.

time educating his people on the dangers of oil exploration which leaves a vast majority of his people impoverished.

Although this essay represents one aspect of my research, it shows the resilience of constituencies (the structure of Movement of the Survival of the Ogoni People – MOSOP)[2], coming together across the kingdoms of Ogoni to create stronger alliances for social change. I argue that the youth wing (National Youth Council of Ogoni People – NYCOP) has created a vibrant space that continues to tell the Ogoni unsavoury story. I discuss how, standing on Ken Saro-Wiwa's template of nonviolent action for justice, the new generation of Ogoni are still seeking their rights within the Nigerian state. This begs the questions as to the place the Ogoni youth have in MOSOP, and their understanding of Ken Saro-Wiwa as the Ogoni hero. As a sympathetic writer from an ethnic minority group (Tiv, Benue State, Central Nigeria) who share similar experience of marginalisation, I cannot claim that my research is politically neutral. But beyond declaring my biases, I was open to other evidences and arguments that contradicted my assumptions.

Thus, I am going to present Ken Saro-Wiwa's narrative strategy. I will explore how his metaphoric interpretation of images and descriptions of life in Ogoniland underscore the horror and brutality of the Nigerian government and Shell very effectively. This, I argue, has made him a hero among the Ogoni youth.

Definition and diversity among Ogoni youth

Contrary to the definition of youth by the United Nations as between 15 and 24 (UN 2007), the situation in most African cultures (Ogoni in this context) has little to do with numbers. What and who falls under the category of youth are those who identify themselves as belonging to a social group and are ready to participate in any defined activity.[3] De Wall

2 Ken Saro-Wiwa established a socio-cultural movement called Movement for the Survival of the Ogoni people – MOSOP in 1990 to demand justice from the Nigerian government for the Ogoni people. As a grassroots organisation and for a fair representation of the entire Ogoni society, MOSOP Comprises 10 sub-bodies: The National Youth Council of Ogoni People (NYCOP), Ogoni Council of Churches (OCC); Council of Ogoni Traditional Rulers (COTRA); Ogoni Students Unions (OSU – secondary schools and below); National Union of Ogoni Students (NUOS- those in tertiary institutions); Ogoni Council of Churches (OCC); Ogoni Teachers Union (OUT); Federation of Ogoni Women Association (FOWA); Ogoni Central Union (OCU); Council of Ogoni Professionals (COP). These are what I referred to above as constituencies.

3 On this issue, it is important to make reference to Ken Saro-Wiwa's relationship with the youth wing of MOSOP at the height of the Ogoni struggle in the early 1990s. He was in his 50s but aligned himself with the youth group who seemed to understand his strategy more than those who considered themselves elders. His education of the young Ogonis reflects in their ongoing determination in the fight for justice.

(UNDP, 2006, p. 16) confirmed this when he said: 'in pre-colonial African societies, adulthood was reserved for men with relative wealth and social status, and a very small number of older women. Everybody else retained the status of minors, however old they were. With colonialism and mission education, the idea of an automatic transition based on age was introduced.' In support of the idea, I agree with Kehily's (2013) remarks that the *age-bound* idea is a Western construct. As most Africans believe, it denies people from contributing their potential positive roles in the society.[4]

Ken Saro-Wiwa founded MOSOP at the age of 50 and, as most literature on Ogoni reports, his radical articulation of Ogoni grievances appealed and still appeals to Ogoni youth who see the Nigerian government as the villain. As Kukah (2011) argued, the struggle "had a David and Goliath symbolism to it." Participants (both male and female) in my research were between the ages of 25-50, which is contrary to the definition of youth here in Europe. They considered themselves as youth, and the reason in this context was based on the analysis that the MOSOP youth wing is a more vibrant force in the fight for justice. Thus, it is important to remember that youth status only reflects a general age grouping.

From the research experience, there was, of course, a great deal of diversity to be found amongst the Ogoni youth population: socioeconomic status, gender, sectionalism (e.g. emphasis or bias towards people from other kingdoms),[5] life-experiences, abilities, constraints and so much more. Hence, the knowledge generated from the research, which is reflected in this essay, represents all youth perspectives rather than a single idea. The multiple voices that were brought into the discussions reflect the diversity of the Ogoni society.

Ken Saro-Wiwa metaphoric narrative: power to Ogoni youth

In this section, I discuss Ken Saro-Wiwa's metaphoric narrative to show how it has succeeded in capturing the moral high ground for youth resilience in Ogoni. About this idea, it is significant to borrow Stone-Medi-

4 As a Catholic priest with vast experience in parish and other pastoral engagements, this idea is explained in practical terms where the Catholic Men Organisation (CMO) finds it difficult in getting members because the majority want to belong to the Catholic Youth Organisation even when their age and status qualifies them for the CMO group. People feel they can contribute more with the youth group. The impression is that belonging to the CMO group is like succumbing to old age, hence, not having the capacity for doing anything reasonable.

5 Ogoniland consists of six kingdoms: Babbe, Leme (Eleme), Gokana, Ken-Khana, Nyo-Khana, and Tee (Tai). However, some members of the Eleme kingdom deny any link to an Ogoni political community and distance themselves from the other Ogoni kingdoms (for more discussion see Ikoro, S.M. 1996. *The Kana language*. Leiden: Research School.).

atore's (2003) argument in *Reading across Borders…* where she emphasised the importance of critical storytelling in the pursuit of mature political consciousness. However, traditional philosophy, as she opines, does not recognise the epistemic value of more subjective, story-like representations of the world, being more concerned about truth and objectivity and preferring the resoluteness it associates with rational thought. Personal stories, she claims, are not mere stories because they can be justified according to their contribution to understanding and critical engagement. Personal stories of struggle challenge basic norms and categories, such as assumptions about identity and agency. Besides, they are not always strictly autobiographical-at least not in a standard sense-but rather are stories about peoples, and about national goals, intending specifically to reclaim and revalue the experience of the oppressed (see Uraizee, 2011).

In his seminal books *Genocide in Nigeria: The Ogoni Tragedy* (1992), *Sozaboy: A Novel in Rotten English* (1985), *On a Darkling Plain: An Account of the Nigerian Civil War* (1989), and *Nigeria: The Brink of Disaster* (1991) Ken Saro-Wiwa creates "story images," or "meaningful content" combined with 'moral, affective, and aesthetic qualities' (Stone-Mediatore 2003, pp. 34-35). He uses rhetorical strategies to tell stories about the suffering that government and the multinational oil companies have inflicted on Ogoni since the 1960s. By using the rhetorical strategies such as *monstrous horror, betrayal, testifying voices*, and *awakening solidarity* (Uraizee, 2011), Saro-Wiwa creates a language of terror and suffering to construct his main argument, which is that oil drilling is devastating taking perspectives of the Ogoni experience into cognisance. He publicises the terrible conditions of life in Ogoni and projects the multinational oil giant, the Shell company as 'the Leviathan to whom we have been forced to surrender all our Rights including our very life' (Saro-Wiwa 1992, p. 63).

According to Stone-Mediatore (2003), by stories or narratives, Ken Saro-Wiwa had in mind 'a pattern of identifiable actors and action-units that are qualified through metaphor and other poetic devices and that are related together through a coherent structure of beginnings and endings' (2003, p.33). While it might seem that such a definition describes any theoretical account-explanations, they bring about possibilities for understanding by allowing readers to 'test and revise their community's taken-for-granted narrative paradigms' (Stone-Mediatore, 2003, p.185) through experience of more marginalised perspectives. Ken Saro-Wiwa's storytelling involves more explicitly moral and political objectives for bringing about constitutional democracy. These are summarised in the Ogoni Bill of Rights (OBR).[6]

Drawing from the above analysis, it is evident that Ken Saro-Wiwa's narratives arose out of a conviction that the Nigerian government have a moral and political responsibility for ending the *genocide* of the Ogoni people and the complete devastation of their environment. There is no doubt that his caustic writing reveals one of his nonviolent campaign strategies, which resonates with the entire Ogoni population. His stories about how the government loots Ogoni resources have become part of the everyday life of the Ogoni people. The intergenerational impacts of oil activities revealed through the narratives is still shining new light on the youth assumptions and practices and scrutinising their effectiveness in achieving long-term social change. There are calls for new ideas and leadership from every corner, with young people often taking the lead on identifying issues as well as creating the space for dialogue and action.

Ogoni Youth: new strategies for social change

Just like in other parts of the world, the Ogoni young people are now leading their movement for change. While in the field, it was interesting to watch and listen to how they were asking fundamental questions about issues (socio-economic and political inequalities) that are affecting their society. Their resolve to continue to challenge the Nigerian state, which they see as 'a predatory institution whose forms of rule, run against their own conceptions of authority and the norms and values that legitimise it' (Okonta 2008, p.6). Though they rely on Ken Saro-Wiwa's nonviolent paradigm, the present strategy is that they are responding to the new exigencies within the wider Nigerian and international community – climate justice articulated in the Ogoni clean-up actions, and gender equality issues within the movement.

Looking back across historic (for information on the rift between Ken Saro-Wiwa and the Ogoni elites see Kukah 2011) and current events within the Ogoni society, especially as it relates to their movement – MOSOP, young activists expressed a lot of fear as regards obstacles (polarisation or factionalisation caused by government interference) to their activism. Despite their interpretation of this moment as weakness in the movement,

6 Considering the Nigerian policy of exclusion that leaves the Ogoni people at the margin of the Nigerian socio-economic life, the Ogoni elites decided and put together their grievances in writing: the Ogoni Bill of Rights (OBR). This was presented to the Nigerian government in 1990. Retrieved from http://www.bebor.org/wp-content/uploads/2012/09/Ogoni-Bill-of-Rights.pdf accessed 31 July 2020

the majority believe the youth power (interpreted as youth resilience) is the most appropriate position to maintain as they move forward seeking justice. Apart from their ability to talk about the violence directed toward the Ogoni, their political vision (Ogoni egalitarianism)[7] is to defeat the hegemonic economic interest of the Nigerian state that has monopolised their environment and resources. They believe that addressing the problem from the root is the best way to creating meaningful and enduring change.

Ken Saro-Wiwa's key role in Ogoni politics brought their struggle to both national and international prominence, thereby broadening the basis for participation. As mentioned earlier, the development of the various constituencies changed 'the power dynamics within MOSOP, and in many respects altered the balance of power and the relation between the leadership, the parent body and these affiliate bodies' (Kukah 2011, p.114). Even though it created, and still creates, an internal strife within MOSOP (especially the relationship between the youth and the current MOSOP leadership) they still value the importance of having charismatic leaders. However, from the Ogoni experience of Ken Saro-Wiwa's gruesome murder by the government to halt the movement,[8] their strategy is against identifying a leader for fear he or she might also be killed.

My interpretation of the current situation as it relates to the young is that they identify their struggles and unite on their common experience. Even though there are issues of generational differences and individuals having their own principles as alluded to above, the majority of young people show unrelenting effort towards the complex intersections in the movement. Youth have also developed new ways of careful cultivation of strong working relationships that bring together the strengths of the movement. The youth coalition is bringing new life and possibility to

7 This is captured in the "property owner rights" frame *Miideekor*, which is used in everyday Ogoni vocabulary and allows every Ogoni to understand the rationale behind their participation in the protest. The word *Miideekor* refers to the palm wine produced in one out of the five workdays of the Ogoni work week, namely *Deemua, Deebom, Deezia, Deezion* and *Deekor*. Traditionally, the palm wine tapper may leep (removal of dirt from the palm tree e.g. dry leaves) the palm wine produced in four out of the five days. Deekor or the first day of the week was a special day for showing appreciation to the landlord. The process of returning the one day (Deekor) a week proceeds due to the property owner is *Miideekor*. Thus, *Miideekor* symbolises the relationship between the owner of a palm field and the palm wine tapper. In this case, Ogoni is the owner of the palm field while the Nigerian government is the palm wine tapper (see Agbonifo, J. (2019) *Environment and Conflict: The Place and Logic of Collective Action in the Niger Delta*, London, Routledge, p.71)

8 In their book *Kill a Leader, Murder a Movement?...* Bob and Nepstad examined the issue of movement repression by governments. Comparing the El Salvadorian liberation and the Ogoni autonomy movements, they explore the risk involved in being a movement leader especially in authoritarian regimes. For more discussion about the effect of such political repression, see Bob, C. & Nepstad, S. E. (2007) 'Kill a leader, murder a movement? Leadership and assassination in social movements', *American Behavioral Scientist*, vol. 50, no. 10, pp. 1370-1394, 10.1177/0002764207300162

the fight for justice, reshaping the ways they organise, and building new forms of facilitative leadership. It is obvious that they are able to employ direct tactics and take risks, act as community messengers to their marginalised communities, make new connections between issues and focus on the movement for the survival of their people.

Conclusion

From the argument established above, it is evident that though the youth in Ogoni are living in turbulent times, it is possible to locate them within the Ogoni movement MOSOP. The essay has demonstrated how the narrative strategy of Ken Saro-Wiwa has helped forge in the youth a consciousness of the unjust and inhumane Nigerian socio-political system. It is this that drives them in their struggle for justice. The Ogoni youth are still able to articulate their grievances within the context of the Ogoni Bill of Rights. Though the essay draws from the wider context of my research, it focused on the point (education of the youth through narrative strategy) that has helped shape the current action of the youth in Ogoni; the reason why Ken Saro-Wiwa is seen as the Ogoni hero.

It is evident that the young Ogonis are seeking a future for themselves and they believe that the "youth power" can lead them to success. They believe that the gateway to development is about local people striving for change, knowing that they have the right to expect greater equality in their own lives, and that there is a pathway to achieve this – through their nonviolent struggle. The Ogoni youth are convinced that the fundamental responsibility of every Ogoni is to act against the regime of rapine (the seizure of another's property) and secure a new inclusive and participatory politics for the development of the Ogoni people in Nigeria.

References

De Wall, A. (2002) 'Realizing Child Rights in Africa: Children, Young People and Leadership', in De Waal, A. and Argenti, N. (eds) *Young Africa: Realizing the Rights of Children and Youth,* Trenton, Africa World Press

Kehily, M. J. (2013) 'Youth as a Construction', in Curran, S., Harrison R. and Mackinnon, D. (eds) *Working with Young People*, 2nd ed, London, SAGE Publications Ltd

Kukah, M. H. (2011) *Witness to Justice: An Insider's Account of Nigeria's Truth Commission*, Ibadan, Bookcraft

Okonta, I. (2008) *When Citizens Revolt: Nigerian Elites, Big Oil and the Ogoni Struggle for Self-Determination*, Trenton, Africa World Press

Saro-Wiwa, K.

(1992) *Genocide in Nigeria: The Ogoni Tragedy,* Port Harcourt, Saros International Publishers

(1991) *Nigeria: The Brink of Disaster*, Port Harcourt, Saros International Publishers.

(1989) *On a Darkling Plain*. Port Harcourt, Saros International

(1985) *Sozaboy: A Novel in Rotten English*, New York, Longmans

Stone-Mediatore, S. (2003) *Reading Across Borders: Storytelling and Knowledges of Resistance* New York, Palgrave Macmillan

United Nations (2007) 'The World Programme of Action for Youth', https://www.un.org/esa/socdev/unyin/document/wpay_text_final.pdf, accessed 13 December 2019.

United Nations Development Programme (UNDP) (2006) Youth and Violent Conflict: Society and Development in Crisis. https://reliefweb.int/sites/reliefweb.int/files/resources/88C-UNDP%20youth.pdf accessed 23 August 2020

Uraizee, J. (2011) 'Combating Ecological Terror: Ken Saro-Wiwa's "Genocide in Nigeria"' The Journal of the Midwest Modern Language Association, vol. 44, no 2, pp.75-91

A People-Driven Non-Violent Revolt

By Nnimmo Bassey

For some of us Ogoni has become the training ground for environmental justice. It has remained the prime territory for learning how difficult it is to undo ecological harm once it has occurred; once it has been allowed to fester and take root. The Ogoni people have also given us a clear base to understudy the workings of a people-driven, non-violent revolt; the challenges, the pitfalls and the triumphs. Ogoni has been a metaphor for ecocide and an inspiration for resistance.

The Ogoni Bill of Rights[1] of November 1990 is a major milestone document, serving to coalesce the pains, dreams and demands of the Ogoni people. It stands as a major decolonial document and was the precursor of similar pursuits by other ethnic nationalities in the Niger Delta, including the Kaiama Declaration of the Ijaws, lkwerre Rescue Charter, Aklaka Declaration for the Egi, the Urhobo Economic Summit Resolution and the Oron Bill of Rights, amongst others.[2]

Article 16 of the Ogoni Bill of rights stated that "neglectful environmental pollution laws and sub-standard inspection techniques of the Federal authorities have led to the complete degradation of the Ogoni environment, turning our homeland into an ecological disaster." Three decades later, this summation remains accurate, even more poignant.

Standing at the centre of the Ogoni experience are a number of personalities one of whom is Ken Saro-Wiwa. His leadership at various levels and platforms left indelible marks on the socio-ecological struggles of the Ogoni people and others. Some of us make regular visits to the polluted sites in Ogoni to remind ourselves that ecocide in any location is a crime against Mother Earth and all our relatives. Ogoni reminds us all that corporate greed can convert a verdant land into a land where humans and other living beings are literally either sick or dead.

The literary output of Ken Saro-Wiwa helped to preserve his thoughts for us and for generations yet unborn. Needless to say, his bluntness also made him controversial. That can be understood because when you are a

1 Ogoni Bill of Rights (1990). http://www.waado.org/nigerdelta/RightsDeclaration/Ogoni.html

2 Nnimmo Bassey. 01 August 2013. Two Years After the UNEP Report – Ogoni Still Groans. http://nnimmo.blogspot.com/2013/08/two-years-after-unep-report-ogoni-still.html

minority fighting to breathe, those whose knees are pressed into your neck would claim that as long as you can complain it means you can breathe. In other words, their knees would only be lifted from your neck when you fall silent. Dead. The noose snuffed the physical life from him 25 years ago, but he still speaks. His satirical story, *Africa Kills Her Sun*[3], shows how fiction can chisel a message in stone. Writing about how a priest would approach to pray for a person about to be executed, he said: "The priest will pray for our souls. But it's not us he should be praying for. He should be praying for the living, for those whose lives are a daily torment."

His fiction was never altogether fictive. According to one Onookome Okome, "These fictive characters are modelled on social types and local events. This explains why some of these characters provoked great and enthusiastic, albeit sometimes acerbic debate in Nigeria's literary history." Okome goes on to say that "his political ideas about the Nigerian Federation were even more controversial. His book on the Nigerian civil war (*On A Darkling Plain: An Account of The Nigerian Civil War*), carefully conceived around the minority/majority problems of Nigeria's ethnic groups, aroused heated hate-debate, especially among members of the three largest Nigeria ethnic groups."[4]

His focus on bringing the plight of the Ogoni people to the world in the context of the unequal majority-minority relations within the Nigerian state, combined with the brutal state capture by notorious transnational oil companies obviously earned him many adversaries, including those who eventually orchestrated his judicial murder along with Barinem Kiobel, Saturday Dobee, Paul Levura, Nordu Eawo, Felix Nuate, Daniel Gbokoo, John Kpuinen and Baribor Bera. Their death was both an epitome of the viciousness of an unholy matrimony between a rapacious transnational entity and an autocratic state, and a glaring failure of international diplomacy.[5]

Saro-Wiwa was conscious of the fact that the consequences of the struggle could be dire, even when prosecuted non-violently. *In Silence Would be Treason: The Last Writings of Ken Saro-Wiwa*, he stated that he signed a death warrant when he "undertook to confront Shell and the Nigerian establishment." He wrote that if his life was not cut short, he would look forward to

3 Ken Saro-Wiwa (1989) *Africa Kills Her Sun*. (short story)
4 Onookome Okome (2000). *Before I am Hanged: Ken Saro-Wiwa – Literature, Politics and Dissent.* Trenton: Africa World Press, Inc.
5 Patrick Naagbanton (2016). *Footprints of Nkpoo Sibara, Dele Giwa and Ken Saro-Wiwa*, Vol. 1. Makurdi: DNA Traeces Empire Limited

"A few more books, maybe, & the opportunity to assist others." In a letter he wrote on 19 June 1995, he stated: "I know they will do everything to resist us and that they may still want me out of the way. I am not careless of my safety, but I do recognize and have always recognized that my cause could lead to death. But as the saying goes, how can man die better/than facing fearful odds/for the ashes of his fathers/and the temple of his Gods? No, one cannot allow the fear of death to dent one's beliefs and actions. I only wish there were more Ogoni people on the ground. However, the cause cannot die."[6]

The matter of having more Ogoni people on the ground to keep the struggle alive remains an active concern; a task that must be done. Yes, the cause has not died, and 27 years after the expulsion of Shell from Ogoni, the oil wells are still not gushing crude. However, the spate of oil pollution remains and the clean-up of the territory although commenced has its speed and mode of delivery highly contested. Having layers of leadership on the ground is essential for any movement. The Ogoni struggle has been kept alive by the deep mobilisations that have gone on over the years and by the clear understanding of the value of their environment and cultural autonomy by the majority of the people. Organisational efforts have floundered and become quite fractious at times, probably due to an alternative notion of sacrifice and superficial commitment to the ideals of the collective. It may well also be driven by impulses of indiscipline and possible conspiracy to subvert the pursuit of the common good.

The Ogoni Bill of Rights spoke of the land turning into an ecological disaster. This position was validated by the United Nations Environment Programme (UNEP) in their report of the Environmental Assessment of Ogoniland.[7] The report, submitted to the Nigerian government in August 2011, revealed extensive pollution of the soil by petroleum hydrocarbons in land areas, swamps and sediments.

In a particularly troubling case, the groundwater at Nisisioken Ogale was found to have an 8cm layer of refined oil floating on it. The drinking water in this community was recorded to have benzene, a carcinogen, at over 900 times the permissible levels according to World Health Organisation guidelines. At some oil well locations, hydrocarbon contamination

6 Íde Corley, Helen Fallon and Laurence Cox, eds (2018). *Silence Would be Treason – Last Writings of Ken Saro-Wiwa*, 2nd ed. Daraja Press – http://mural.maynoothuniversity.ie/10161/

7 UNEP (2011). Environmental Assessment of Ogoniland. https://www.unenvironment.org/explore-topics/disasters-conflicts/where-we-work/nigeria/environmental-assessment-ogoniland-report

was found to be at least 1000 times above Nigerian drinking water standards. Sadly, the people continue to drink such contaminated water due to lack of alternatives. Soil investigations by UNEP revealed that hydrocarbon pollution had seeped down to a depth of 5 metres at some places.

Our government's response to the UNEP report was the setting up in 2012 of an initial Hydrocarbons Pollution Restoration Project (HYPREP) under the Ministry of Petroleum Resources which is a key player in the pollution of Ogoniland. This project was later on revamped and renamed the Hydrocarbons Pollution Remediation Project (HYPREP) in 2016 because it was saddled with the task of remediating the pollution, not restoring it. This name change would likely have made a good subject for a satirical piece by Saro-Wiwa, had he witnessed its creation.

At the time of this writing, HYPREP reported[8] the completion of work at seven remediation sites out of 21 lots that it had awarded contracts for. It also announced that they were in the process of awarding contracts for an additional 26 lots. The clean-up process has been dodged with complaints over the speed of the works as well as over the calibre of contractors assigned the tasks. There have also been contentions over whether the remediation should have commenced with the most complex sites or the simpler ones.

The sorest point is that of non-provision of potable water in the affected communities, many years down the remediation road even though it was considered an emergency measure by both the UNEP and the Ogoni people. HYPREP officials inform that work will soon start in rehabilitating six moribund water supply works across Ogoniland.

A recent visit[9] to some of the remediation sites by this writer was quite revealing. Whereas the depth of hydrocarbon pollution was at an alarming 5 metres at the time UNEP conducted its study, the state of affairs has deteriorated over the years. Hydrocarbons pollution was found to have now gone as deep as an alarming 10 metres at Lot 2. One other finding was that 30,000 litres of petrol were recovered from this Lot. We saw a layer of hydrocarbons on the excavated pit at Lot 16, at Korokoro community, beside the tanks of recovered crude that were stored nearby.

The recovery of crude oil from the remediation sites show that without the remediation, the pollution would obviously sink deeper, leaving the

8 Michael Simire (10 September 2020) Legislators express satisfaction with pace of Ogoniland clean-up amid concerns. https://www.environewsnigeria.com/legislators-express-satisfaction-with-pace-of-ogoniland-clean-up-amid-concerns/

9 This visit was on Friday 11 September 2020

disaster more intractable. It also offers a stark warning to oilfield communities that even where the land looks normal, tests need to be done at intervals of time to ensure the integrity of what lies beneath the surface.

November 1990, when the Ogoni Bill of Rights was issued, and November 1995, when Ken Saro-Wiwa and the other leaders were executed, are cardinal milestones in the march for ecological and socio-political justice for the Ogoni people and all marginalised peoples that are victims of destructive extractivism.

Twenty-five years after the judicial murders, the wounds inflicted on the Ogoni people are yet to heal. Twenty-five years after the act, the Nigerian State has still not found the place to formally exonerate the Ogoni leaders and foster healing in the land. Twenty-five years after the macabre act, even the sculpture in honour of the Ogoni 9 lies captive at the Apapa quays in Lagos, Nigeria, held by a system that is afraid to come to terms with an artistic artefact.[10] Who will tell the Nigerian government that arresting and detaining a piece of sculpture in an effort to block the memory of crimes committed by the state is an exercise in futility?

Ken Saro-Wiwa saw it all. He felt it. He told it. He challenged all. His last public speech or allocutus, stands like a banner at the head of a marching column and we do well to pay attention:

> *We all stand before history. I am a man of peace, of ideas. Appalled by the denigrating poverty of my people who live on a richly endowed land, distressed by their political marginalization and economic strangulation, angered by the devastation of their land, their ultimate heritage, anxious to preserve their right to life and to a decent living, and determined to usher to this country as a whole a fair and just democratic system which protects everyone and every ethnic group and gives us all a valid claim to human civilization, I have devoted my intellectual and material resources, my very life, to a cause in which I have total belief and from which I cannot be blackmailed or intimidated.*[11]

10 Susanna Rustin (5 November 2015). Ken Saro-Wiwa memorial art bus denied entry to Nigeria. https://www.theguardian.com/environment/2015/nov/05/ken-saro-wiwa-memorial-art-bus-denied-entry-to-nigeri

11 Ken Saro-Wiwa (1995). Trial Speech of Ken Saro-Wiwa. https://en.wikisource.org/wiki/Trial_Speech_of_Ken_Saro-Wiwa

'The Voice of the Spirit of Ogoni': A Brief Exploration of The Niger Delta Cultural Landscape with Reference to the Ogoni

Abayomi Ogunsanya

Introduction

Nigeria's Niger Delta became firmly stamped on the consciousness – or maybe the *conscience* – of the world during the concentration of events that climaxed with the execution of the environmental activist, writer, and businessman Ken Saro-Wiwa and eight others (The Ogoni Nine) on November 10, 1995, by Nigeria's military junta. Prior to that momentous event, there was not a lot, at least in the popular press, reported about the region, except for the occasional mention when Nigeria was discussed in relation to oil or during political upheaval.

The Ogoni, one of the eleven minority ethnic groups in the region, received even less attention until events forced the world to take notice in the 1990s. Since then, and following renewed agitation for environmental justice in the region in recent years, the Niger Delta, and Ogoni in particular, has become a major reference point in any discussion about contemporary Nigeria. As might be expected, given how oil has shaped the contemporary history of Nigeria, much of what is discussed about the region in the media pertains to the political economy of oil, and the violence that is done to the people and their land. There is little discussion on the rich culture of the area in popular discussions, although academic literature on the region is abundant.

This essay briefly explores some of the cultural practices that define the Niger Delta region with particular focus on the Ogoni ethnic group. The essay does not pretend to be encyclopaedic in coverage; rather it attempts to provide some context to the Ogoni story. This context is often missed in discussions about the Niger Delta, as the hotbed of criminal violence and various intrigues connected with oil exploration and the concomitant environmental degradation.

Geography of the Niger Delta

In geographical terms, the Niger Delta encompasses about 20,000 km^2 in landmass and is said to be one of the largest wetlands in the world. It is

> a vast floodplain built up by the accumulation of sedimentary deposits washed down the Niger and Benue rivers. It is composed of 4 main ecological zones: coastal barrier islands, mangroves, freshwater swamp forests, and lowland rainforests. The mangrove forest area extends for 500 km from the mouth of Benin River in the west to the Imo River in the east. Along with a short strand coast in the Calabar to Rio del Rey estuary, mangrove vegetation extends over an area of 9000 km^2. This represents the largest mangrove forest in Africa.[1]

A significant portion of the Niger Delta region is made up of a network of creeks and small islands, 'making it very difficult to navigate and to establish large settlements.'[2]

As a geopolitical entity, the Niger Delta is recognized as the south-south zone in Nigeria and comprises Akwa Ibom, Bayelsa, Cross River, Delta, Edo, and Rivers State. Ogoni is in Rivers State. An extended portion of the Niger Delta includes: Abia, Imo, and Ondo – all oil producing states.

As an ethnic zone, the Niger Delta comprises Ogoni, Andonis, Calabaris, Edos, Efiks, Ibibios, Ijaws, Isokos, Itsekiris, Opobos, and Urhobos, each asserting separate socio-cultural identities but united in living in a part of the world where oil is found in super abundance.

Ogoni occupies 'a minute parcel of land of about 404 square miles'[3], contains a sizeable percent of Nigeria's total oil reserve, and is home to close to one million people. According to several historical and ethnographic accounts, the Ogoni were culturally conservative, 'fiercely independent' people who often 'regard foreigners with a healthy dose of suspicion.' Here is how Karl Maier describes them:

> 'As long as their land, streams, and mangrove swamps could provide them sustenance, the Ogoni people wanted as little as possible to do with the outside world. As a tiny ethnic unit of six kingdoms,

1 Ajao, E.A. & Anurigwo, Sam (2002). 'Land-Based Sources of Pollution in the Niger Delta, Nigeria.' *Ambio*, Vol. 31, No. 5, 442-445

2 Francis, P., LaPin, D., Rossiasco, P. (2011) Securing Development and Peace in the Niger Delta: A Social and Conflict Analysis for Change. Woodrow Wilson International Center for Scholars.

3 Maier, K. (2000). *This House has fallen: Nigeria in Crisis.* Colorado: Westview Press

> they jealously guarded their culture and identity by maintaining strict prohibitions on intermarriage with most of their neighbours. Their harsh environment literally served as an umbrella protecting them from the hostile intentions of outsiders. In the nineteenth century the dense forests were the perfect refuge from the slave trade, which used the nearby Bonny and Imo Rivers to transport the human cargo to the sea and the waiting ships for the onward journey to the New World. Most Ogonis retreated to the less accessible reaches of the Ogoni plain, where they made their living through agriculture, fishing, and hunting.'4

Tales of Origin

In prehistoric times, the Ogoni, like most ethnic groups with no archaeological and written records of their specific origins, was not a single ethnic formation. In tracing the earliest roots of Ogoni ethnic identity, Isumonah (2004) identifies two types of myths on the origin of the Ogoni, namely, autochthony and migration, and observes that "the former credits the founding of the Ogoni to another realm and the latter, which is more credible with some empirical evidence, regards Nama as the Ogoni's ancestral cradle land."[5]

Forging Cultural Identity

During the course of territorial expansion through annexation of villages, conquests of other peoples, and trading, Ogoni cultural identity began to be forged from the amalgamation of social and cultural practices among the different peoples that would later make up the Ogoni Kingdom. Isumonah (2004) gives a sweeping description of some of the social and cultural beliefs and practices among the people at that time:

> 'The various groups enjoyed autonomy in parallel through similar social, economic and political institutions and cultural and religious practices, at times seeking symbolic recognition from the mythical progenitor at Nama. Notable commonalities were the 'customs of protecting villages by deep trenches and the use of five-day week, unlike the four- and eight-day weeks of surrounding Ibibio and Ibo clans,' as well as matrilineality....The people used common currencies (*Giaradaa, Nama-kpugi* and *Ka-kpugi*) and

4 Ibid, p. 82
5 Isumonah, V. Adefemi, 'The Making of the Ogoni Ethnic Group.' Africa (pre-2011); 2004; 74, 3; pg. 433.

> markets for local and international trade. They developed mortgaging and property valuation systems run by expert agents *(Pya tedee)*, *bureaux de change* run by the *Pya Nyaa Na kpugi* and business norms and class differentiation by wealth....The Yaa rite of passage from youth into responsible adulthood served as an instrument of both social stratification and political recruitment of the virgin male youth. The *Kpan-kpaan* secret societies was an important political force and law enforcement agency.'[6]

As Ogoni cultural identity became more distinctly recognizable over the next few centuries and the people came to identify themselves as an ethnic group within the Niger Delta cultural archipelago, many of the beliefs and practices described above during prehistoric times became parts of the people's identity. The Ogoni have retained these beliefs and practices and often deploy them in shaping their engagements with their neighbours and while confronting some of the challenges of the modern world. For example, Maier (2000) reports that during the struggle for environmental justice in Ogoniland, Ken Saro-Wiwa, the champion of that struggle, "saw himself as the Wiayor, a mythical Ogoni character who comes down from heaven to liberate his people."[7] Similarly, the Ogoni cause itself is sometimes conceptualized in spiritual terms and Saro-Wiwa saw himself as an embodiment of that cause. What is called the Voice of the Spirit of Ogoni is "a fetish god central to the Ogoni belief system" and Saro-Wiwa once explained how he became an embodiment of that Spirit: "One night in late 1989, as I sat in my study working on a new book, I received a call to put myself, my abilities, my resources, so carefully nurtured over the years, at the feet of the Ogoni people and similar dispossessed, dispirited and disappearing peoples in Nigeria and elsewhere."[8]

If Ogoni people are often described as fearless and fiercely independent, it is simply because many of them, like their kinsman Ken Saro-Wiwa, often put themselves in the service of the Voice of the Spirit of Ogoni. Isaac Adaka Boro, Ken Saro-Wiwa's forerunner in the Ogoni struggle, was a case in point; though an Ijaw, he was an example of one of those who allowed the Spirit to possess them and became an inspiration for other Niger Delta fighters demanding justice for their people. When Ogoni people resisted the onslaught of British troops for a long time, despite their land

6 Ibid, p. 437
7 Maier, K. *This House has Fallen*, p. 92
8 Ibid, p.92

being a British protectorate, and were only subdued in 1914, "the same year that Nigeria came into existence as a single entity", it was because they relied on the Spirit.

Ken Saro-Wiwa spelt out clearly the role of the supreme goddess Bari in *Genocide in Nigeria*.[9] This is also discussed by Agbonifo (2019)[10] in relation to the Ogoni struggle. Both the Christian God and the Ogoni deities provide moral incentives to contention. Kpone-Tonwe's PhD thesis[11] also provides valuable insights.

9 Saro-Wiwa, Ken. (1992) *Genocide in Nigeria: The Ogoni Tragedy*, Port Harcourt, Saros International Publishers

10 Agbonifo, John (2020). *Environment and Conflict: The Place and Logic of Collective Action in the Niger Delta*. London: Routledge

11 Kpone-Tonwe, Sonpie (1987). "The Historical Tradition of Ogoni, Nigeria. PhD thesis submitted to School of Oriental and African Studies (SOAS), University of London. https://eprints.soas.ac.uk/28874/1/10673043.pdf

Amnesty International and the cleanup of the Niger Delta

Mark Dummett

Since Shell first discovered oil near the village of Oloibiri in 1956, the Niger Delta has become Africa's most valuable oil-producing region, and the Anglo-Dutch corporation has prospered enormously.

Shell does not publish a breakdown of its earnings by country, but Reuters estimated that it had earned 4 billion US dollars from oil and gas production in Nigeria in 2017, which was around 7 percent of its total global output.[1]

Shell's former chief economist and prominent British politician Sir Vince Cable laid out the historic importance of Shell's Nigeria business in his memoirs, describing it as the "jewel in the crown of the exploration and production division, the company's elite corps."[2]

But while money has flowed to Shell's headquarters in The Hague and London, its operations have undoubtedly come at the cost of the human rights of people living close to the oil fields of the Niger Delta.

Numerous oil spills a year from poorly-maintained pipelines and wells, along with inadequate clean-up practices, continue to damage the health and livelihoods of the Niger Delta's many inhabitants, who largely remain stuck in poverty.

Data from Shell's own spill incident reports[3] reveal that from 2011-18 the company reported a huge number of spills – 1,010 – along the network of pipelines and wells that it operates. Spills have a variety of causes – from third-party tampering, to operational faults and corrosion of aged facilities. Shell blames most spills on theft and pipeline sabotage, rather than its own negligence.

But research by Amnesty International[4] and our Port Harcourt-based

1 Bousso, R. (2018) 'In Nigeria, Shell's onshore roots still run deep', Reuters, available online: https://www.reuters.com/article/us-nigeria-shell-insight/in-nigeria-shells-onshore-roots-still-run-deep-idUSKCN1M3069.

2 Cable, V. (2010) *Free Radical: A Memoir*, London: Atlantic Books.

3 Shell (2019) 'Spill prevention and response in Nigeria', *Shell Sustainability Report 2019*, Available online: https://reports.shell.com/sustainability-report/2019/special-reports/spill-prevention-and-response-in-nigeria.html

4 Amnesty International (2017) *Investigate Shell for complicity in murder, rape and torture*, [online] 28 November 2017, Available at: https://www.amnesty.org/en/latest/news/2017/11/investigate-shell-for-complicity-in-murder-rape-and-torture/.
Amnesty International (2009) *Nigeria: Petroleum, Pollution and Poverty in the Niger Delta*, [online]

partners at the Centre for the Environment, Human Rights and Development (CEHRD)[5] has shown that the company's facts and figures emerge from a flawed process for identifying the volume, cause and impact of oil spills. The research has also exposed that this process often lacks both independence and oversight, partly because the government regulators are so weak. As a result, Shell's findings cannot be trusted.

For example, the company's report for a spill in the Bodo area of Ogoniland in 2008 claimed that only 1,640 barrels of oil were spilled. However, based on an independent assessment, Amnesty International calculated that the total actually exceeded 100,000 barrels.[6] Shell defended its far lower figure for years, but in November 2014 during a court case in the UK, Shell was finally forced to admit that the amount was indeed larger than it had previously stated.[7]

Even in cases of spills caused by theft or sabotage, Shell still has responsibility. Nigerian law requires all pipeline operators to employ the best available technology and practice standards. These include measures to protect against spills resulting from third party interference, such as by strengthening or burying pipelines and surveillance. Nigerian law also requires oil companies to clean up spills from their infrastructure, regardless of the cause.

Internal company documents and other sources collated by Amnesty International, prove that Shell staff have known for years that underinvestment, poor maintenance or equipment failure have been a major cause of the spills.

For example, in 1994, the head of environmental studies for Shell Nigeria, Bopp Van Dessel, resigned, complaining that he felt unable to defend the company's environmental record, "without losing his personal integrity."[8]

---

30 June 2009, Available at: https://www.amnesty.org/en/documents/afr44/017/2009/en/.
Amnesty International (2013) *Nigeria: Bad information: Oil spill investigations in the Niger Delta*, [online] 7 November 2013, Available at: https://www.amnesty.org/en/documents/afr44/028/2013/en

5 Centre for Environment, Human Rights and Development (2015) *Polluted Promises: How Shell failed to clean up Ogoniland*, [online] 3 September 2015, Available at: https://cehrd.org.ng/downloads/polluted-promise-briefings-2014.pdf, p.10.

6 Amnesty International (2012) *Shell's wildly inaccurate reporting of Niger Delta oil spill exposed*, [online] 23 April 2012, Available at: https://www.amnesty.org/en/latest/news/2012/04/shell-s-wildly-inaccurate-reporting-niger-delta-oil-spill-exposed/.

7 Amnesty International (2014) *Court documents expose Shell's false claims on Nigeria oil spills*, [online] 13 November 2014, Available at: https://www.amnesty.org/en/latest/news/2014/11/court-documents-expose-shell-s-false-claims-nigeria-oil-spills/

8 Chatterjee, P. (1996) 'Ex-Shell Environment Chief for Nigeria Blows Whistle,' *Inter Press Service*, [online] 18 May 1996, Available at: http://www.ipsnews.net/1996/05/environment-ex-shell-environment-chief-for-nigeria-blows-whistle/.

The same year an internal paper revealed that Shell had not properly funded its pipelines and other infrastructure in Nigeria: "One measure of this deterioration is the frequency and severity of oil pollution incidents caused by corrosion and other integrity failures in the production system."

In 1996, a Shell Nigeria "Country Business Plan" identified that its "infrastructure [was] poorly designed and maintained."[9]

In 2002, an internal Shell presentation stated: "the remaining life of most of the [Shell] Oil Trunk Lines is more or less non-existent or short, while some sections contain major risk and hazard."

In 2008, a US diplomatic cable stated that a contractor with many years' experience of laying pipelines in the Niger Delta reported that, "73 per cent of all pipelines there are more than a decade overdue for replacement. In many cases, pipelines with a technical life of 15 years are still in use thirty years after installation."

And in 2009, a Shell employee warned in an email that: "[the company] is corporately exposed as the pipelines in Ogoniland have not been maintained properly or integrity assessed for over 15 years."

Shell's negative impact was of course first, and most effectively, put under the spotlight in the 1990s by Ken Saro-Wiwa, and the Movement for the Survival of the Ogoni People (MOSOP). They argued that while outsiders had grown rich on the oil that was pumped from under their soil, pollution from oil spills and gas flaring had, "led to the complete degradation of the Ogoni environment, turning our homeland into an ecological disaster."

Shell is not the only one at fault. Nigeria's regulation of the oil industry is undoubtedly weak and lacks independence. Government agencies responsible for industry regulation and enforcement are under-resourced, ineffective and in some cases compromised by conflicts of interest. Nigeria's courts have failed to offer the victims of human rights abuses a meaningful avenue for seeking justice for the oil spills which have blighted the Niger Delta, as well as the lives and livelihoods of its communities.

The failures of the courts have resulted in Shell avoiding being held effectively to account in Nigeria. It has benefitted enormously from extracting oil in a context where there is little or no government oversight and no effective safeguards.

As elaborated in globally endorsed standards like the UN Guiding Prin-

9 Amnesty International (2020) *On Trial: Shell in Nigeria: Legal actions against the oil multinational,* [online] 10 February 2020, Available at: https://www.amnesty.org/en/documents/afr44/1698/2020/en/, p. 18.

ciples on Business and Human Rights[10], companies have a responsibility to respect human rights wherever they operate in the world. To meet that responsibility, companies should take steps to prevent, identify, address and account for their human rights impacts. This includes remediating any harm they have caused or contributed to.

The responsibility to respect human rights exists independently of a state's ability or willingness to fulfil its own human rights obligations. So, if a state where a company operates, such as Nigeria, is unable or unwilling to enforce applicable laws to protect human rights from abuse, the company must still act to ensure respect for human rights in their operations. Shell itself expressly says that it uses the UN Guiding Principles to inform its global approach to human rights.

However, the pollution in the Niger Delta has had a significant human rights impact. Shell has consistently failed to fulfil its responsibility to respect the human rights of communities there.

Mark Dummett was keynote speaker at the Maynooth University 2017 Ken Saro-Wiwa Seminar. His presentation is available on the Maynooth University YouTube channel.

10 United Nations (2011) *Guiding Principles on Business and Human Rights*, [online] 16 June 2011, Available at: https://www.ohchr.org/documents/publications/guidingprinciplesbusinesshr_en.pdf, p.13.

Why Ken Saro-Wiwa matters for climate justice

Laurence Cox

As a social movements specialist I often find myself talking to nice, well-meaning students and professionals in the global North. Often they are (rightly) focussed on the terrifying reality of climate crisis and desperate to know what to do – but the strategies for change that are easy to find turn out to be very simplistic, shallow to the point of being trivial, and completely inadequate to the scale of the problem.

In particular, many of the forms of action they are presented with ignore the history of what has actually worked in ecological movements – in their own countries in previous decades, or around the world at the moment. We are offered solutions that suit us, whether or not they actually have any track record of winning against the huge concentrations of power and wealth, and the entrenched cultural and social habits, that underpin carbon capitalism.

When I can, I tell them some of the story of Ken Saro-Wiwa and MOSOP as a way of helping them start to think more seriously, in ways that might actually work. The Ogoni are one of the world's most disadvantaged populations – so rural and remote from the centres of power that even their exact numbers are uncertain – and yet they were able to effectively resist Shell, which on recent figures is the world's 18th largest economic entity, bigger than the economies of Mexico, Sweden or Russia, in times of a military dictatorship. That alone suggests that we should try to learn from them.

We need to think more about movements

We are used to researching problems and issues, and sometimes to digging into the structural reasons behind them – but for various reasons the journalists, teachers and charities who do this often find it much harder to discuss honestly how to tackle those problems – which is a "political" question. And so concerned citizens are brought to see the problems, but rarely get to learn from the experience of organising and mobilising strategies around these kinds of problems.

A major reason for this is that the climate crisis is "baked in" to a breakneck capitalism which depends on permanent growth and within which the industries that most contribute to global heating – fossil fuels, air and car transport, agribusiness and so on – are very powerful. It is one thing to discuss policy solutions that can conceivably be adopted by today's governments without upsetting those economic interests. It is another thing entirely to ask what it means if the survival of our societies depends on breaking the power of those industries and transforming the whole direction of state policies.

And so awareness is raised about problems, but the entrenched, systemic nature of those problems is rarely addressed – leaving the people whose awareness has been raised in a state of anxiety or despair, keen to find things that look like simple solutions, and ignorant of the long history of attempts to actually tackle these problems.

How words lose their meaning

"Climate justice" is just the latest round of attempts to name the shape of this problem. The phrase draws from "environmental justice" as an organising strategy – which emphasised the impossibility of tackling the root causes of ecological destruction without challenging the economic system that gives rise both to a ruined planet and to ruined lives and communities, and so highlighted the need to build strong alliances between ecological movements and popular struggles for social justice.

For exactly the same reasons, a problem on the scale of climate change requires these kinds of alliances in order to build an effective social majority – whether Saro-Wiwa's mobilising of a desperately poor rural population to resist Shell, or in a very different kind of society the Norwegian "just transition" campaign, which brings environmentalists, trade unionists and church bodies together around the demand for climate jobs.

Yet already, in Ireland and much of the world, policy makers and NGOs have simply substituted the phrase for earlier wordings on global warming or climate campaigning, so that it means precisely no change to their activities. When I was a student in another century, the same thing was happening to the phrase "sustainable development". This once meant "we have to think how we can bring together large-scale improvements in ordinary people's lives around the world without relying on permanent economic growth", but by that time it had become watered down to the point of meaning almost nothing, either in terms of the goal or of how to get there.

Earlier phrases with a similar meaning – such as "eco-socialism" or "red-green" – have proved less easy to absorb into business as usual, but regularly get forgotten and are rediscovered every decade or so.

This process of noticing, and then forgetting, has been going on at least since the 1970s. In that past fifty years, the inability of our societies to look this problem clearly in the eye, understand the systematic reasons why things keep getting worse, and to learn from past attempts to tackle them, has cost us all dearly. We have very little time left in which to indulge our desire to find easy solutions to difficult problems.

Why climate justice matters

Irish newspapers, and social media, are full of comments which pit the needs of "ordinary people" against the supposedly elitist concerns of environmentalists. Environmentalists who act in high-handed ways, allying themselves with the rich and powerful and blaming the poor for the problem, play right into this particular perception.

If climate crisis is an outcome of hugely damaging industries, in turn deeply entrenched within the policy process and states' priorities, and ultimately expressing a capitalism which is determined to pursue infinite growth on a finite planet ... then the rich and powerful will not, and cannot, save us. It is by challenging their priorities and pushing for a radically different kind of economy – which in turn means huge changes in society, politics and culture – that we can hope to avert utter disaster.

But if we want to challenge the fossil fuel giants, the airline companies, the meat industry, the political parties and the financial institutions that are driving us towards destruction ... ecologists are going to need allies, from those who are neither powerful nor rich. Without a mass movement, we are not going to win. And that movement (or rather alliance of movements) cannot be brought together without involving the needs of the poor and powerless for a more just world.

This is true both in terms of what is needed now to create the social majorities that are required to face down the determined opposition of those whose livelihoods, power and status depend on keeping the carbon show on the road – and in terms of what a new society might look like in the future. It is not credible that we could create a genuinely "sustainable" way of living – one that enables us to survive ecologically and that is socially and politically stable – unless it meets the needs of the large majority of people on this planet.

Perhaps by now it is clear why these sorts of considerations are not centre-stage in the strategies presented to school students, newspaper readers and NGO members.

It may also be clear why it is worth remembering Ken Saro-Wiwa.

What works?

Last summer I had a number of discussions with excited members of the Extinction Rebellion Group[1] who had just come into possession of "a little knowledge" – a misreading of research that suggested a magic number of protestors that would inevitably bring about social change. Almost none of them – thoughtful, educated people – had any idea that there was actually a history of struggles against climate change that could be learnt from.

The practical part of that history is above all one of indigenous resistance to the extraction and transport of fossil fuels, to drilling and pipelines. It is a history of some of the poorest and most oppressed people on the planet, managing to stand up to, and sometimes defeat, ruthless states and mega-corporations. In recent years this has been very visible in North America, as First Nations and Native Americans have mobilised again and again against tar sands projects, gas pipelines and so on. Images from Standing Rock or the Wet'suwet'en protests have gone round the world.

Saro-Wiwa's effective struggle against Shell similarly mobilised the poorest of the poor against huge odds: an indigenous population against one of the world's largest corporations, backed up by a military dictatorship that was willing to execute the movement's leadership and unleash brutal terror against Ogoni villages. And it was broadly successful: Shell remain persona non grata in Ogoni to this day, while the dictatorship has fallen (in part also due to the Ogoni struggle).

If we think about ecological survival without any awareness of these stories, it is very tempting to think that if it was only possible to tone down the conflict, get policy-makers on board, get an issue into schools, convince journalists ... then it could all be solved. In that perspective, conflict is just a personality flaw, not in any way inherent in (say) how fossil fuel corporations make their money.

From this perspective, social justice – and alliances with movements of the poor and oppressed – are unnecessary and awkward add-ons to a

1 https://extinctionrebellion.uk/

simple environmental "message" that would slip down much more easily if only all its rough corners were rubbed off.

But as we can see in Ogoniland, the real story is the opposite of this. In order to challenge Shell and the military dictatorship effectively, Saro-Wiwa had to mobilise the large majority of a desperately poor and downtrodden population. The 60% (never mind 3.5%) of Ogoni who are commonly said to have taken part in the 1993 protests, and the large majority of Ogoni who still identify with Saro-Wiwa's MOSOP, took a lot of convincing.

A social majority for climate justice

Their needs and concerns – economic justice and social development, indigenous self-determination and human rights – are not awkward add-ons to a simple environmental "message". They are precisely the things that made it possible to mobilise effectively against one of the world's largest companies and a murderous, corrupt regime.

It is worth remembering that Saro-Wiwa was a key player in pushing the concept of "indigenous" onto the UN's agenda, as a way of furthering this struggle. When, today, Native Americans or First Nations resist the fossil fuel industry and the US or Canadian states as indigenous populations, they are drawing on this shared history.

If we want to bring together social majorities for a world we can all live in, we need to look to the Niger Delta and to Ken Saro-Wiwa, as well of course as to other indigenous struggles against fossil fuels around the world where we see small and massively disadvantaged populations, hugely vulnerable to state violence, face off against and often defeat the forces that are driving us all to destruction.

We also need to have the humility and the good faith to learn from people who are so far away from us – in their lives as well as geographically – that we need to go the extra mile in terms of finding out where and how we can listen to them. That means not being satisfied with the very first answers we find, but asking critical questions about whose experiences and movements those answers are actually based on – and reaching out towards learning from the struggles of the world's poor and powerless.

If we are able, at this eleventh hour of human civilisation, to mobilise social majorities for a world that we can all live in, we will need to pay more attention to Ken Saro-Wiwa.

The Centrality of Culture in the thinking of Cabral and Saro-Wiwa

By Firoze Manji

I want to share some thoughts about the commonalities between Amilcar Cabral, the Guinea-Bissau revolutionary[1], and Ken Saro-Wiwa, especially in relation to the centrality of culture in the struggle for freedom.[2]

The struggles Cabral led against Portuguese colonialism contributed to the collapse not only of Portugal's African empire, but also precipitated the Portuguese revolution of 1974/5 and the downfall of the fascist dictatorship in Portugal, events that he was not to witness as he was assassinated in 1973.

Cabral and Saro-Wiwa were separated by two eras, the one involving the struggle for independence in Africa, the other dealing with the consequences of the failures of independence and the rise of neoliberalism. There were continuities between the two eras. "Cabral and Saro-Wiwa sit together in this transformative and unfinished space," wrote Laurence Cox, "asking questions that remain important in Ireland as in Africa."[3]

Despite this separation, they had much in common.

Both sought self-determination for their people. Both were clear that self-determination, not secession, was what they were fighting for. Self-determination and secession are often confused and considered synonymous. Self-determination is about the struggle for justice, dignity and an attempt to establish an inclusive universalist humanity, whereas secession is by definition an act of exclusion, defining the self through the exclusion of the other.

The tragedy for Saro-Wiwa was that the struggle for self-determination for the Ogoni came in the wake of the Biafran war of secession, the leadership of which Saro-Wiwa was highly critical. The struggle of the Ogoni people for self-determination could easily come to be seen as a continuation of that secessionist movement, despite Saro-Wiwa's insis-

1 Amilcar Cabral was the founder and leader of the Guinea-Bissau and Cabo Verde liberation movement, Partido Africano da Independência da Guiné e Cabo Verde (PAIGC). He was a revolutionary, humanist, agronomist, poet, military strategist, and prolific writer on revolutionary theory, culture and liberation.

2 This essay is based on the keynote address from the annual Ken Saro-Wiwa seminar, Maynooth University Library, Maynooth University, 15 November 2018

3 https://www.irishtimes.com/culture/books/the-legacy-of-ken-saro-wiwa-at-maynooth-university-1.3730031 (Accessed 24 June 2020)

tence against secession (although he was sometimes ambiguous about the distinction).

While Cabral and Saro-Wiwa were clearly exceptional individuals, it was to the movements in which they were involved, and which they helped to create, that the credit must go, for organising and for endeavouring to give birth to a new world.

We often characterise movements such as those led by Cabral and Saro-Wiwa as being expressions of resistance. But I think they are more than that. Far from being the 'resistance', they fought to give birth to a new world.[4] Their attempts were met by fierce resistance by the powerful whose interests lay in maintaining their status and the status quo. We need to insist that it is they who are the resistance. It was the Portuguese colonial regime and the Abacha neocolonial regime in collaboration with Shell that were the resistance to the efforts of the movements that PAIGC (African Party for the Independence of Guinea and Cape Verde) and MOSOP (Movement for the Survival of the Ogoni People) sought to birth.

Giving birth is always an act involving the struggle to overcome the violence of resistance. This is as true of a seedling emerging from the ground as it is for the child being born. Genuine movements for freedom never chose the path of violence, but they almost always face the violence of those that resist the birth of the world they are seeking to deliver. We are seeing precisely the same issues emerging in the Black rebellion that erupted in the USA (and beyond) following the torture and killing of George Floyd. The struggle to give birth to Ireland was met with fierce, violent and terrorist resistance by the British state. There was no choice but to endeavor to defend it. But more importantly, it is community organizing, that is the basis of defence in the case of Ireland as it was in the case of Guinea-Bissau and Nigeria.

So, let us agree: The state and corporations, not us, are the resistance.

In Guinea Bissau, PAIGC had created liberated zones that, at the time of Cabral's assassination, covered some two-thirds of the country. There, completely new structures of popular democracy were established in which peasants were the decision makers. The Portuguese currency was banned, and a system of barter exchange was established in its stead. Women played leading roles in political decision making. The rekindling of culture and

4 I am indebted to Michelle Alexander's essay: "We are not the resistance", https://www.nytimes.com/2018/09/21/opinion/sunday/resistance-kavanaugh-trump-protest.html (Accessed 24 June 2020)

pride in their own histories, languages, stories and music flourished. New health, education and other services were established. They were creating a new world. But they had no choice but to ensure the movement had the means to defend the new society that had been built. PAIGC politics was not about promoting violence, but about defending the birth of a new society from the genocidal violence of Portuguese colonialism.[5]

Both Cabral (at the hands of his own comrades, those who were to become the neo-colonial rulers of the future) and Saro-Wiwa (at the hands of the neocolonial Abacha regime) paid the ultimate sacrifice for their audacity to both think and create, in their time, a new world. This is what distinguishes them from so many others: it was not only having a dream that another world was possible, but also having the courage to create that world in the present. It was that which presented such a threat to those who resist new births.

I make this point because it is in the crucible of the struggle for a new world that real culture evolves as a weapon of liberation.

> To be able to subject millions of humans to the barbarism of enslavement, slavery and colonial domination required defining them as non-humans or less than humans, required attempts to condemn African people as being less than human. That process required a systematic and institutionalised attempt at the destruction of existing cultures, languages, histories and capacities to produce, organise, tell stories, invent, love, make music, sing songs, make poetry, produce art, philosophise, and to formulate in their minds that which they imagine before giving it concrete form, all things that make a people human.[6]

This attempt to destroy the culture of Africans, points out Cabral, turned out to be a signal failure. For, while colonialism destroyed the institutions on the continent, the memories of their culture, institutions, art forms, music and all that which is associated with being human remained both on the continent and in the diaspora where the enslaved Africans found themselves. The enslavers, the slave owners, and all those who profited from these horrors, including the emerging capitalist classes of

5 See Firoze Manji and Bill Fletcher Jr (eds) (2013): *Claim No Easy Victories: The Legacy of Amilcar Cabral.* Dakar: Daraja Press & CODESRIA

6 Manji F (2019): Emancipation, Freedom or Taxonomy? What Does It Mean to be African? In Vishwas Satgar (ed): *Racism After Apartheid: Challenges for Marxism and Anti-racism.* Wits University Press. Pp 49-74

Europe, engaged in a systematic recasting of human beings as non-humans or lesser beings, a process in which the Christian church and the European intelligentsia were deeply involved.[7]

Whatever the material aspects of domination, 'it can be maintained only by the permanent and organized repression of the cultural life of the people concerned' wrote Cabral. The use of violence to dominate a people is 'above all, to take up arms to destroy, or at least neutralize and to paralyze their cultural life. For as long as part of that people have a cultural life, foreign domination cannot be assured of its perpetuation.'[8]

For Saro-Wiwa:

> 'The advent of British colonialism was to shatter Ogoni society and inflict on us a backwardness from which we are still struggling to escape. It was British colonialism which forced alien administrative structures on us and herded us into the domestic colonialism of Nigeria. ... As a result of domestic colonialism, the Ogoni people have virtually lost pride in themselves and their ability, have voted for the multiplicity of parties in elections, have regarded themselves as perpetual clients of other ethnic groups and have come to think that there is nowhere else to go but down... Yes, we merely exist; barely exist.'[9]

Culture, wrote Cabral, is

> 'the product of ... history just as a flower is the product of a plant. Like history, or because it is history, culture has as its material base the level of the productive forces and the mode of production. Culture plunges its roots into the physical reality of the environmental humus in which it develops, and reflects the organic nature of the society.

Culture, insists Cabral, is intimately linked to the struggle for freedom. While culture comprises many aspects, it

> 'grows deeper through the people's struggle, and not through songs, poems or folklore. ... One cannot expect African culture to advance unless one contributes realistically to the creation of

7 Manji F (2017): Culture, power and resistance: Reflections on the ideas of Amilcar Cabral. In *State of Power 2017*. Amsterdam: Transnational Institute. https://www.tni.org/en/publication/state-of-power-2017 (Accessed 24 June 2020)
8 Cabral, A (1979): *Unity and Struggle: Speeches and Writings* New York: Monthly Review Press, pp139-40
9 Saro-Wiwa, Ken (1995). *A Month and a Day and Letters*. UK: Ayebia Clark Publishing Limited, p50

> the conditions necessary for this culture, i.e. the liberation of the continent.'[10]

In other words, culture is not static and unchangeable, it advances only through engagement in the struggle for freedom.

In this, he echoes Frantz Fanon:

> To fight for national culture first of all means fighting for the liberation of the nation, the tangible matrix from which culture can grow. One cannot divorce the combat for culture from the people's struggle for liberation... national culture takes form and shape during the fight, in prison, facing the guillotine and in the capture and destruction of the French military positions. ... National culture is no folklore ... [it] is the collective thought process of a people to describe, justify, and extol the actions whereby they have joined forces and remain strong.[11]

Ken Saro-Wiwa's identity as a member of the Ogoni people, along with his political activism is inseparable from the content of his novels. 'The writer cannot be a mere storyteller,' writes Saro-Wiwa, 'he cannot merely x-ray societies weaknesses, its ills, its perils. He or she must be actively involved shaping its present and its future.'

> The most important thing for me, is that I've used my talents as a writer to enable the Ogoni people to confront their tormentors. I was not able to do it as a politician or a businessman. My writing did it. And it sure makes me feel good! I'm mentally prepared for the worst, but hopeful for the best. I think I have the moral victory.[12]

Saro-Wiwa believed that 'literature in a critical situation such as Nigeria's cannot be divorced from politics. Indeed, literature must serve society by steeping itself in politics, by intervention, and writers must not write merely to amuse... They must play an interventionist role.'[13]

Saro-Wiwa, like Cabral before him, believed that the writer 'must take part in mass organisations' and 'establish direct contact with the people.'[14]

---

10 Cabral (1979) op ci

11 Fanon, Frantz (1967) *The Wretched of the Earth*. Hammondsworth: Penguin.

12 Letter to William Boyd (2004). Review of Ken Saro-Wiwa a month and a day: A detention dairy. Ibadan: Spectrum Books Limited

13 Ibid p55

14 Ibid p55

'What they (the authorities) cannot stand, is that a writer should additionally give voice to the voiceless and organise them for action. In short, they do not want literature on the streets! And that is where, in Africa, it must be.'[15]

As to language, Saro-Wiwa commented:

> Furthermore, I have examined myself very closely to see how writing or reading in English has colonised my mind. I am, I find, as Ogoni as ever. I am enmeshed in Ogoni culture. I devour Ogoni food. I sing Ogoni songs. I danced to Ogoni music. And I find the best in the Ogoni world-view as engaging as anywhere anything else. I'm anxious to see the Ogoni establish themselves in Nigeria and make their contribution to world civilisation. I myself am contributing to Ogoni life as fully, and possibly even more effectively than those of Ogoni who do not speak and write English. The fact that I appreciate Shakespeare, Dickens, Chaucer, Hemingway, et al., the fact that I know something of European civilisation, its history and philosophy, the fact that I enjoy Mozart and Beethoven – is this a colonisation of my mind? I cannot exactly complain about it ... Historically, the Ogoni people have always been fierce and independent. They have been known to display an exceptional achievement in their original, abstract masks. As storytellers and in other forms of art the Ogoni are gifted and hold their own easily. The Ogoni have made contributions of the first order to modern African literature in English.[16]

The implicit appeal to a universalist and inclusive humanity is clear in these statements. Cabral had no hesitation in writing for a wider public in Portuguese, but he was insistent that in order to learn from the peasantry, the ability to converse in their languages was needed. 'We must put the interests of our people higher, in the context of the interests of mankind in general' wrote Cabral, 'and then we can put them in the context of the interests of Africa in general.'[17]

'We must have the courage to state this clearly', he said, 'No one should think that the culture of Africa, what is really African and so must be pre-

15 Saro-Wiwa, Ken letter to William Benson, 17 July 1995.

16 Saro-Wiwa, Ken. "The language of African literature: a writer's testimony. (The Language Question)" Research in African Literatures, Spring, 1992, Vol.23(1), p.153(5)

17 Cabral, A. (1979). Unity and Struggle: Speeches and Writings. Translated by M. Wolfers. New York: Monthly Review Press. p80

served for all time, for us to be Africans, is our weakness in the face of nature.'[18]

'Indeed,' says Saro-Wiwa,

> literature must serve society by steeping itself in politics, by intervention, writers must not merely write to amuse or to take a bemused, critical look at society. They must play an interventionist role. My experience has been that African governments can ignore writers, taking comfort in the fact that only few can read and write, and that those who read find little time for the luxury of literary consumption beyond the need to pass examinations based on set texts. Therefore, the writer must be *l'homme engagé:* The intellectual man of action.[19]
>
> He must take part in mass organisations. He must establish direct contact with the people and resort to the strength of African literature – oratory in the tongue. For the world's power and more powerful is it when expressed in common currency. That is why a writer who takes part in mass organisations will deliver his message more effectively than one who writes, waiting for time to work its literary wonders.[20]

'A reconversion of minds – of mental set – is thus indispensable to the true integration of people into the liberation movement,' wrote Cabral. 'Such reconversion – re-Africanization, in our case – may take place before the struggle, but it is complete only during the course of the struggle, through daily contact with the popular masses in the communion of sacrifice required by the struggle.'[21]

As the writings of both Cabral and Saro-Wiwa show, culture is not a mere artefact or expression of aesthetics, custom or tradition, Rather, it is a means by which people assert their opposition to domination, a means to proclaim and invent their humanity, a means to assert agency and the capacity to make history. In a word, culture is one of the fundamental tools of the struggle for emancipation.

18 Cabral, A. (1979). Unity and Struggle: Speeches and Writings. Translated by M. Wolfers. New York: Monthly Review Press. p80
19 Saro-Wiwa (1979) op cit p55
20 Ibid
21 Cabral, A. 1973. *Return to the Source: Selected Speeches of Amilcar Cabral.* New York: Monthly Review Press. p45

Solidarity with human rights defenders against corporate human rights abuse

By Siobhán Curran

Introduction

In the early 1990s, Trócaire[1], an Irish development organisation, campaigned for the release of Ken Saro-Wiwa and eight other human rights defenders who had been arrested and sentenced to death for campaigning for their human rights, in the face of exploitation and environmental damage by Shell. Over 25 years later, indigenous, environmental and land defenders who are resisting unsustainable actions by corporations and states, continue to face killings, violence and intimidation. The most recent report by Global Witness (2020) reports that the highest number of killings of land and environmental defenders in a single year stands at 212, an average of more than four murdered people murdered a week.

Ogoni Struggle

The injustice facing the Ogoni people provoked empathy and condemnation in Ireland. Between the 1960s and the 1990s, an estimated $30 billion had been extracted from Ogoniland, yet the people who lived here were not benefitting from profits generated (Maye, 2010). Shell caused major environmental damage in the area, prompting the formation of the Movement for the Survival of the Ogoni People (MOSOP) who organised in defence of their very survival. Trócaire called on the Nigerian government to release the leaders and ran a full page advertisement in Irish newspapers entitled 'Ken Saro-Wiwa's only crime was to campaign for his people. Now the Nigerian military regime wants to kill him for it.' It noted that '37 years after Shell had begun drilling for oil in the area, 400 square miles of Ogoniland were dotted with oil spills, contaminated water and gas flames.' (Irish Times, 4 November 1995). In 1995, Ken Saro-Wiwa and his eight colleagues (the Ogoni Nine) were executed, causing international shock and outrage.

1 Trócaire works with local partners to support communities in over 20 developing countries with a focus on food and resource rights, women's empowerment and humanitarian response. https://www.trocaire.org/

After these killings, Shell continued to rely on the state military to stifle opposition and to terrorise communities, while they produced billions in profits, at the expense of the Ogoni people and the environment. To this day, MOSOP continues this struggle against corporate impunity and calls for Shell to clean up the effects of its operations, while Shell continues to avoid responsibility and fight these cases through the courts. The struggle of the Ogoni people against corporate harm is replicated throughout the world, and as with the case of the Ogoni people, it is often indigenous communities on the frontlines of this struggle.

Corporate and State Power

Instead of the urgent action that is required to address climate change, which is driven by fossil fuel extraction including oil, some states and corporations continue to act in the pursuit of profit at the expense of human rights and the planet. The communities that Trócaire works with are experiencing the devastating impacts of unchecked corporate and state power, including the displacement of communities to make way for corporate developments, violent evictions, the pollution of land and the complete destruction of livelihoods. Such corporate-related activities are diminishing access to land and water, vital for growing food, shelter and a whole range of interdependent human rights, including access to education and health. This is why Trócaire has prioritised the campaign for corporate accountability, in solidarity with those resisting corporate and state power.

Human Rights Defenders

Those resisting structures of power are doing so in the most challenging of circumstances. Global Witness (2020) notes that the most dangerous sectors have been agribusiness, oil, gas and mining, which are also the sectors driving climate change. In addition to risk of killings and disappearances, human rights defenders are being subjected to undue criminal prosecution and judicial harassment, including criminal charges, arbitrary arrests, detentions, and strategic lawsuits brought by companies. Women human rights defenders working in this context also face gendered risks, which exploit existing inequalities and perceptions about their role in society. The UN Special Rapporteur on the situation of human rights defenders (2019) notes that attacks on women human rights defenders include attacks on their honour and reputation, public shaming, sexual violence, and threats against their children and loved ones. Front Line Defenders

(2020) have highlighted that online smear campaigns, trolling and defamation are regularly used to intimidate shame or harass women human rights defenders. Women are increasingly under attack, not only for what they are challenging but also for who they are.

Bertha Zúniga Cáceres stands beside a mural of her mother, murdered human rights activist, Berta Cáceres. Bertha is the general coordinator of the Civic Council of Popular and Indigenous Organizations of Honduras (COPINH).

Juana Zuniga is treasurer of the Environmental Committee of Guapinol, Honduras. She is a member of the Committee in Defence of Common and Public Assets, and mother of three.

The Business and Human Rights Resource Centre (2020) notes that between 2015 and 2018, 12 carbon majors (active fossil fuel producers that are the largest corporate contributors to greenhouse gas emissions) brought at least 24 lawsuits against 71 environmental & human rights defenders for a total $904m of damages. Trócaire works in Honduras, where community members in the small town of Guapinol are facing years of jail for trying to protect their river from a mining company, which was damaged and dried up when the company started work. This rural town has now been heavily militarised, with constant checkpoints, and intimidation of community members. Eight community members have been imprisoned since September 2019 without trial for their peaceful resistance. The community organization has been labelled an 'illicit and criminal organization'. Those defending human rights are being criminalised, while the perpetrators of abuses and environmental damage can continue to operate.

Global Human Rights Treaty

At the core of this problem is the continued reliance on voluntary commitments to encourage corporations to respect human rights and the environment. There is no binding international legal framework to establish the liability of transnational corporations with respect to human rights and the environment, or to ensure justice for affected communities. The growth of large transnational corporations, with major revenues, lobbying power and influence, and operating across states, poses major accountability challenges. Some governments are unable or unwilling to enforce human rights regulations in relation to the activities of corporations, and at times perpetrate human rights violations themselves, in order to keep or attract investment. This is why communities throughout the world are calling for a global human rights treaty to prevent corporate harm, particularly the activities of transnational corporations who can hide behind complicated corporate structures to avoid accountability. Current business models that result in the destruction of communities, killings, smear campaigns and judicial harassment of human rights defenders, while harming the environment, need to be transformed and this transformation needs to be systematic.

Global Solidarities

Just as Ken Saro-Wiwa and MOSOP fought and continue to fight a battle against corporate impunity, state power and racism, indigenous groups worldwide continue to bear the brunt of this global injustice – often protecting common and natural resources, which are vital to our survival. Advocacy, campaigning and support for a UN global human rights treaty and legislation at regional and national levels, is an act of solidarity with human rights defenders who are challenging corporate harm. It is listening to their analysis of the need for structural, systematic change and transformational change. For too long we have relied on corporations to voluntarily change their practices, while communities on the front lines face forced evictions, pollution of their lands, loss of livelihoods, attacks, smear campaigns and arbitrary detention. Yet, the resistance of communities and movements in response to corporate power continues, despite the risks that these brave people face. We should change this approach and insist on legally binding regulation – we have to protect the land, environmental and indigenous defenders, who are already protecting all of us.

Business and Human Rights Resource Centre (2020) 'Human Rights Defenders and Business: January 2020 Snapshot', https://dispatches.business-humanrights.org/hrd-january-2020/index.html (accessed 29 August 2020)

Front Line Defenders (2020) Global Analysis 2019.

https://www.frontlinedefenders.org/en/resource-publication/global-analysis-2019 (accessed 29 August 2020)

Human Rights Council (2019) 'Situation of women human rights defenders – Report of the Special Rapporteur on the situation of human rights defenders', A/HRC/40/60.

https://digitallibrary.un.org/record/1663970 (accessed 29 August 2020)

Maye, B. (2010) The Search for Justice: Trócaire – A History, Dublin, Veritas Publications

Ken, We Heard You: *Irish Global Solidarity in One Hundred Objects*

By Tony Daly and Ciara Regan

Introduction

Five days before the execution of the poet, writer and environmental activist Ken Saro-Wiwa and his eight colleagues (the Ogoni 9) in Nigeria on the 10th of November 1995, a mock hanging took place outside the Nigerian embassy in Dublin. In the days that followed, there were calls for the expulsion of the ambassador, calls for fact finding missions led by Irish politicians and human rights campaigners to be given travel visas to enter Nigeria and calls for boycotts of Shell, the company at the centre of the upheaval in Ogoniland. In the weeks that followed, young people signed and sent thousands of postcards to Shell's office in Ireland, calling on the company to distance itself publicly from the Nigerian government, as part of learning about the issues through applied-education.

On the first anniversary of the execution of the Ogoni 9, on November 11, 1996, "Ogoni awareness exercises" were staged at Shell garages in Dublin, Galway, Cork, Donegal and Belfast at 12.30pm. Campaigners were urged on by Saro-Wiwa's daughter, Noo, the previous night in Temple Bar.

In the years that followed, many moments would manifest where Ken's actions and words were active in the thoughts and actions of people across Irish society. Moments such as when the Body Shop and Amnesty International organised to cover print and postage costs for members of the public to send free Christmas cards in 1996 listing the names of the children of the 19 Ogoni members in detention since 1994.[1] Members of the volunteer-led Ogoni Solidarity Ireland campaign group joining anti-racism groups on a Day of Action Against Racism and Deportations in May 1999. Moments such as when a group of people changed street signs on Adelaide Road in 2008 to read 'Ken Saro Wiwa Street' where Shell's Irish head office was located. One participant commented:

1 Russell, M. (1996) Saro Wiwa's daughter urges Shell boycott, *Irish Times*, 11 November, https://www.irishtimes.com/news/saro-wiwa-s-daughter-urges-shell-boycott-1.104699, accessed 4 September 2020.

> "It seems fitting to honour Ken Saro Wiwa this way... it is important that his struggle against the destruction of the Niger Delta is remembered by those who deal with energy companies like Shell, in Ireland as well as other places".[2]

Then there were the regular calendar moments too, such as organising a public forum in Wexford for environmental groups in 2010 to both remember the Ogoni Nine and to provide a space to consider how the Ogoni experiences could influence thinking on the Shell-led Corrib Gas project in North Mayo.[3]

Irish Global Solidarity in *One Hundred Objects*

This essay introduces the exhibition project led by the developmenteducation.ie consortium that culminated in a public showcase in Dublin in February 2020. This exhibition included letters and poems written by Ken Saro-Wiwa. One of these poems is discussed. Three lessons learned from this public education experiment and the role of memory and remembering are discussed.

One of the first challenges for the exhibition was to decide what objects could be included to represent these activities (and many others) as part of a broader focus on solidarity activity from Ireland. Underwriting this groundswell of actions across and between groups and people in Ireland and Nigeria was an exchange of letters, of friendship and solidarity between two people: Sister Majella McCarron and Ken Saro-Wiwa and we quickly decided this material must be included.

Introducing the exhibition

'Is ar scáth a chéile a mhaireann na daoine'
We all live in each other's shadow[4]

While the concept of the 'global' is relatively recent, its realities have deep roots. The international stage is but one backdrop against which the story of Ireland has been written. This story – of a people, a culture, a state and an imagination – has been forged not just in Ireland but also in the wider

2 Indymedia.ie (2008) 11 November, 'Street signs changed to honour Ken Saro Wiwa and Ogoni Eight', http://www.indymedia.ie/article/89812, accessed 4 September 2020.
3 Shell to Sea (2010) 'The Annual Ken Saro-Wiwa Memorial Seminar in Wexford', http://www.shelltosea.com/content/annual-ken-saro-wiwa-memorial-seminar-wexford, accessed 4 September 2020.
4 Ken Saro-Wiwa Letters description card, (February 2020) Irish Global Solidarity in *One Hundred Objects* exhibition guide.

world. In all dimensions of that wider world, Ireland has had a presence. Nowhere is that more obvious than in the search for justice and human rights in the world.

Showcasing a people's heritage of global citizenship from Ireland, *Irish Global Solidarity in 100 Objects* uncovered stories and campaigns articulated through artifacts or objects as well as letters and visual arts that bring these histories to life. Based on the fruits of a year-long open call for submissions, the project offered a glimpse into key moments in the story of Ireland's relationship with global human rights challenges over a 50-year period with a focus on a diverse range of actors, institutions and people and campaigns and issues at different stage – not just the 'photo finish' of leaders.

In the early stages of the project we benefitted greatly from the advice of experts in the fields of art, activism and exhibition curating. The story of the exhibition was forged through these encounters, which included meeting Ruth Hegarty of the Royal Irish Academy, teacher educators Fiona King and Tony Murphy in the National College of Art and Design and Anne Kelly of the NCAD Gallery. Support from project partners including Aidlink, Concern Worldwide, Trócaire, Self Help Africa, the Irish Development Education Association, 80:20 Educating and Acting for a Better World and Irish Aid secured the time and resources to support a temporary installation.

Occupying a pop-up space on the former Greene's Booksellers property in central Dublin served the dual purpose of acting as a well-known focal point and cultivating a historically relevant (and apt) space for learning. Attended by c.900 attendees, the five-day pop-up event during Fairtrade Fortnight 2020 and the One Step Up Adult Learning Festival provided a snapshot of Irish engagement with global culture, politics and social issues over the past 50 years. From the film reels by the Holy Ghost missionaries documenting conflict on the ground in Biafra in the 1960s to anti-apartheid Dunnes Stores strike posters, education campaigns on HIV and AIDS and County Wicklow declaring a climate emergency these are but 100 examples from a much broader and deeper story of engagement from Ireland that amplifies, as Seamus Heaney wrote about the Universal Declaration of Human Rights, "the still, small voice" of conscience in a world of increasing human rights 'wrongs', extreme inequalities and unjust practices.[5]

5 Heaney, S. (2008) 'Seamus Heaney on human rights – you should read this…', Amnesty International UK blog, 15 April, https://www.amnesty.org.uk/blogs/belfast-and-beyond/seamus-heaney-human-rights-you-should-read, accessed 4 September 2020.

At the centre of the exhibition was an observation about human flourishing in the world of the neighbourhood, the school or college, the factory or place of work. The exhibition reminded participants and visitors of a shared heritage of engaging in uncomfortable struggles from Ireland in bringing these objects together – not just to be viewed in isolation.

Object exhibits included the lead Climate Strike placard used in the May 24th Student's Strike for the Climate event in September 2019; the University College Dublin Student's Union 'Killer Coke' campus-wide boycott referendum leaflet from 2003; the 'Drop the Debt' debt cancellation collaborative quilt produced in 1999 and Richie O'Shea's biomass-fired cooking stove – overall winner of the 2010 Youth Scientists Exhibition, among many others.

Saro-Wiwa as Poet

Three letters written to Sister Majella McCarron by Ken Saro-Wiwa from military detention were included in the exhibition. These and poems were loaned by Maynooth University Library.[6]

A striking poem from the collection, the poem 'For Sister Majella McCarron', contains many elements that chime with the meaning and purpose of 'global solidarity'. Saro-Wiwa playfully suspends geographical boundaries and distances as he speaks to his 'soul sister' about journeys, ideals, and 'oceans of ink' as they reach out to each other's causes and people, 'Of your Ogoni, my Fermanagh'. The agony that they share through common causes and struggles, as individuals and together, reveals an exchange and a recognition of experiences and feelings infused with empathy and hope.

For Sister Majella McCarron
Sister M, my sweet soul sister,
What is it, I often ask, unites
County Fermanagh and Ogoni?
Ah, well, it must be the agony,
The hunger for justice and peace
Which married our memories
To a journey of faith.
How many hours have we shared
And what oceans of ink poured

6 Maynooth University Ken Saro-Wiwa Libguide, https://nuim.libguides.com/ken-saro-wiwa-collection, accessed 4 September 2020.

From fearful hearts beating together
For the voiceless of the earth!
Now, separated by the mighty ocean
And strange lands, we pour forth
Prayers, purpose and pride
Laud the integrity of ideals
Hopefully reach out to the grassroots
Of your Ogoni, my Fermanagh.

Saro-Wiwa's prison letters reinforce 'the hunger for justice and peace'. The letters solder justice and human rights approaches in their exchange; an essential element of what became an international campaign in the struggle for a remedy in holding decision makers, government officials and a transnational corporation directly accountable. In 'For Sister Majella McCarron' the humanity behind Saro-Wiwa's words permeated the broader campaigns and actions from Ireland.

'...*the hunger for justice and peace which married our memories..*'

In the short time (five days) the exhibition was running, we had a constant flow of visitors, from those who planned to visit and those who happened to be passing by. On one of the days, we met someone who had been passing by and decided to come in to have a look around. To his surprise, he found a black and white photograph of his wife on a picket line. The photograph was taken by photographer Derek Speirs in 1985, who had visited the exhibition two days prior to this and the photograph was part of a collection of posters and photos called 'Dunnes Stores Strikers on the Picket Line'. She was one of the Dunnes Stores strikers. It had brought back a flood of memories for him, beaming with the pride he had for her.[7]

Dozens of stories such as this appeared as the objects arrived. These are important stories from the people involved – in many different yet essential ways – that perhaps would have gone untold, lost in time. This became an early goal for us in organising it the exhibition: to remember. To remember and engage with our activist heritage and our impact in the world around us. Right now, more than ever, cultivating an interest in and actively engaging with memory projects is vital. While Mandela famously said 'history depends on who wrote it', equally famously, Churchill argued

7 Speirs, D. (1985) Dunnes Stores Strikers on the Picket Line, photograph

'history will be kind to me, for I intend to write it'. Memory just isn't about historical fact but also impact and affect.

Narratives for Discussion

As fraudulent varieties of facts are dispersed into the bloodstream of the internet on an hourly basis providing opportunities to 'prove' that events happened and evidence exists to verify the claim are vital. As part of engaging in cultural activities the exhibition was a place for learning, and unlearning, based on multiple perspectives and narratives on display.

Pablo de Greiff, Special rapporteur on the promotion of truth, justice reparation and guarantees of non-reoccurrence carried a sentiment shared by others on a panel discussion on history, teaching and memorialisation processes in May 2014, stating

> "States should keep in mind that their options were not either to forget or to remember the past but rather to discuss what kind of public space should be made available for remembering the past and for allowing the expression of a plurality of views."[8]

Extending this position in 2017, de Greiff stated:

> '...in periods in which opinions in the media, including social media, are increasingly seen to be untethered from any factual basis, the effective use of archival documentation can prevent the manipulation of 'memory' to instigate conflict'[9]

Pascale R. Bos' recent work on memory, the Holocaust and the after effects of 'history' in an American context is worth reminding people about:

> 'There are echoes everywhere of fascism – propaganda, misleading the public, scapegoating used to be historical topics and now all of these echo in the present. This is why it is important to continue to learn about it and know about it.'[10]

8 United Nations High Commissioner for Human Rights (2014) Summary of panel discussion on history teaching and memorialization processes, United Nations, A/HRC/28/3.

9 Special Rapporteur on the promotion of truth, justice, reparation and guarantees of non-recurrence (2017) Promotion of truth, justice, reparation and guarantee of non-reoccurrence report for the 72nd UN General Assembly, United Nations, A/72/43099.

10 Bos, P.R. (2020) 'Do We Understand How The Holocaust Happened? With Pascale R. Bos Part 1', Getting Curious with Jonathan Van Ness podcast, 22 April, https://www.jonathanvanness.com/gettingcurious/episode/3427199e/do-we-understand-how-the-holocaust-happened-with-pascale-r-bos-part-1, accessed 14 June 2020.

Taken together, memorialisation processes such as the *100 Objects* exhibition are vital spaces for engagement, storytelling and learning multiple social histories and perspectives.

Three Lessons Learned

A number of brief reflections have stuck out for those involved. Here, we expand on three highlights.

1. There was a huge response – and expressed need – for global solidarity public memory projects

The range of actors, institutions, voices and supporters were vastly more diverse, broad and interesting than official stories will ever tell. The *100 Objects* project presented, first and foremost, a plurality of voices and experiences that have underwritten a diverse range of globally-focused issues and campaigns from Ireland. This 'people's history', populated by volunteers, youth groups, teachers, trade union stewards and photographers, for example, illustrates the depth of previous campaigns that are not typically accessible on a first reading of official reports of these stories. A perfect example of this is the power of letter writing as evidenced by the letters from Ken Saro-Wiwa to Sister Majella.

Taken together, the 100 objects paint a complicated and deep picture of actions and activities from Ireland as a sample of achievement and struggle from the shop to the classrooms and to the petrol pump.

2. Spaces to encounter materials from previous solidarity campaigns are few and should be cultivated/protected

In a digital age, the memory of activism and campaigns can be short when viewed through a digital lens alone. Decades of debates, documents, agit-prop exercises and learning resources surrounding campaigns are buried in boxes or adorning walls or in cupboards as keep-sakes. The need to cultivate opportunities for reflection based on real-world struggles is vital as many themes of the remain – environmental defenders and the extraction of natural resources, or unjust trade – or are reinvented. There is a wealth of experience and depth to be gained from offline interventions with physical objects acting as both 'proof' and a stimulus for further engagement.

Furthermore, 'transformative education' has been mainstreamed in recent years as part of 'high level' agendas in the United Nations context, now spoken of as a stock phrase by policy makers and Agenda 2030 for Sustainable Development acolytes.[11] Creating opportunities to provoke critical engagement and supply informal curricula are hugely valuable enterprises when looking to historical artefacts and the search for implications for new audiences (and learners) such as the *100 Objects*. They offer pathways for curiosity, evidence for further exploration and study that, in a period flooded with false and misleading 'information', can act as evidence of what happened, who was involved and what outcomes were reached for learners and teachers today.

3. **There are important lessons for educators and activists that objects from previous campaigns can tell today when seen together, if we dare to heed them.**

Stories of change and the struggles for a better world have a lot to offer activists and activism today. During an era where the explosion of fake news, false or misleading tweets, memes, 'reports' and more, proof has never been more important. Creating opportunities to be curious, and discover the legacies and actions of others, such as what has been happening through a project like the *Irish Global Solidarity in 100 Objects* exhibition, provides evidence of what has happened during an era of slippery truths and a yearning for trustworthy signposts.

Conclusion

This project only scratched the surface of movements packed with thousands of encounters that underpin creative resistance, civil disobedience and educational approaches to encourage critical change in a highly unequal world.

Expanding our sense of 'the curriculum' – both formal and nonformal – is at the heart of a project like this. Critically appraising successes, failures and public interest / engagement also invites an active 'reading' of history that can, even 25 years after the execution of Ken Saro-Wiwa and the

11 The 2030 Agenda for Sustainable Development, adopted by all heads of state and government at a special United Nations summit in 2015, provides a shared blueprint for peace and prosperity for people and the planet, now and into the future. At its heart are the 17 Sustainable Development Goals (SDGs), which are an urgent call for action by all countries - developed and developing - in a global partnership to eradicate poverty and achieve sustainable development for all by 2030. Irish diplomats and officials played in active role in the development of the SDGs as co-chair of the final intergovernmental negotiations. Organisations and groups in Irish civil society participated in the largest international consultation on the formation of the Goals through the 'World We Want' platform, co-hosted in Ireland by The Wheel and Dóchas. The SDGs built on the previous set of development goals, the eight Millennium Development Goals, which guided global action on the reduction of extreme poverty in its various dimensions from 2000-2015. For more, see The Sustainable Development Goals Agenda, United Nations, https://www.un.org/sustainabledevelopment/development-agenda.

other Ogoni 8 peace activists, draws a thread of connections from climate strikers to Shell to Sea flotilla kayakers to Nestlé's changing position on Fairtrade. The production of leaflets, organising annual events and tours for members of Ken Saro-Wiwa's family and members of the Movement for the Survival of the Ogoni People (MOSOP) also left a distinctive footprint on education and awareness raising activities in Ireland.

A project like this forces the organisers, the public and participants to think critically and to question their place and role in the world, and invites us to think about what object each of us might add, taking on HIV and AIDS educator and activist Michael J Kelly's advice:

> 'The challenge for each and every one of us is: can I be satisfied that it should remain so? If not, then what am I going to do about it?'[12]

Not everything to be learned is 'online'. Get outside get into 'everyday activism' with others and make your own discoveries through action.

The *Irish Global Solidarity in One Hundred Objects* collection was digitised and launched in September 2020 with supporting teaching materials on the developmenteducation.ie website.[13]

12 Kelly, M.J. (2010) Foreword, in Duffy, V. and Regan, C. (ed) *This is What Has Happened: HIV and AIDS, women and vulnerability in Zambia*, 80:20 Educating and Acting for a Better World.

13 *Irish Global Solidarity in 100 Hundred Objects* exhibition (2020), developmenteducation.ie, https://developmenteducation.ie/100objects, accessed 8 September 2020.

Open Access and the struggle for justice: The Maynooth University Ken Saro-Wiwa Collection

Firoze Manji and Helen Fallon

Publishing in its finest form is not merely the creation of a product for sale, distribution and consumption. It should be considered as an essential cultural process of encouraging collective reflection, thinking, deepening interactions, as well as stimulating organising for justice and dignity. It should enable us to enhance our understanding of the perspectives of different peoples, populations and classes, whether through narrative, literary, analytical, or artistic forms. Given the developments of new technologies of production and dissemination, there are immense possibilities for achieving this. But there are equally significant obstacles.

Publishing is dominated by a small number of large companies. In line with what has become a generalised phenomenon in the era of neoliberal capitalism, companies involved with publishing, especially of academic materials, have experienced a significant concentration. For example, the top five most prolific publishers account for more than 50% of all papers published in 2013, and 70% of papers in the social sciences (Larivière V., Haustein S., Mongeon P., 2015). The major publishing houses, based in the global North, are deriving substantial profits. For example, the UK-based media group, Pearson, reported a revenue in excess of US$7000 million in 2015 (Collyer, F.M., 2018). Most libraries in the global South are simply unable to pay the inflated prices for books and subscriptions to journals and magazines.

The profits made by publishing companies have escalated, while at the same time the actual cost of production – typesetting, printing, and diffusion – has declined, as the fabrication process is made much easier. The fact that much of the information that is published is based on research the public has already paid for (either directly through grants or indirectly through the provision of public education), makes the situation even more irksome and unjust. In the neoliberal era, the putative efficiency of the private sector is frequently lauded; but it is the substantial subsidy received by the companies from the public purse that makes their business appear efficacious.

The commodification of knowledge production and the monopolisation of publication has strongly affected academic practices, influencing the choice of content and even the creative processes required for the production of art and knowledge. Monopolisation allows subscription rates to be inflated, making costs particularly prohibitive for those in the global South (Collyer, F. M., 2018). But most importantly, commodification affects what is considered legitimate: the experiences of the global South, and in particular those of the "wretched of the earth", have little or no exposure or influence on public discourse. Consequently, a Eurocentric and elite perspective dominates.

Moreover, the inheritance of the colonial relationship in academia is often perpetuated today in ways that parallel how the extractive industries (mining and agricultural) operate in the global South: they source raw materials/'primary products' which are then processed in the global North. So, in academia, we see how data and information are collected from the global South as raw materials for articles, books and theses that are then processed/published in the North. A glance at most journals will show how few articles refer to intellectual work undertaken in the global South.

While the emergence of open access initiatives is welcome, it does not necessarily resolve the problems faced by researchers and writers in the South. Materials published in open access journals are often paid for by the authors or their institutions, a system that effectively closes the door for those who are either unable or cannot afford to pay to have their work published. Some journals do make selected articles available on open access and libraries internationally are striving to make more content available through institutional and subject repositories. Sites such as Open Doar (Directory of Open Access Repositories), Directory of Open Access Journals, Core Open Access Research Papers and the Social Science Research Network are valuable, but there are many other resources that incur prohibitive subscription costs. An informative interview with an Ethiopian researcher on the challenges of open access in Africa, sheds light on issues researchers in the Global South face (Cochrane, L. and Lemma, M.D, 2019).

The situation is perhaps worse in the case of books. While the cost of producing and printing books has dropped significantly over the last decade, this has not always been reflected in the changes in the retail price. The cost of printing in much of Africa, for example, is exorbitant both because of the predominance of the use of older technologies, but also because paper has to be imported. What is surprising is that the retail price of e-books is often little different from the price of the printed equivalent. While reproducing printed copies involves labour, costly technologies and paper, the cost of the production and reproduction of electronic books is much less. The establishment of sites such as Directory of Open Access Books and Libgen that make available electronic versions of books is a life-saver for many, especially those who work towards building societies based on justice, freedom and dignity.

In the current period of growing impoverishment of populations across the world associated with neoliberalism, finding ways to encourage open access to poetry, novels, music, song, art, literature, philosophy and all forms of publications is a challenge. However, open access offers potential opportunities for making the voices of the dispossessed accessible and is an intrinsic component of the discourse on freedom and justice. Publishing for the commons in this context can be a powerful act of solidarity, making knowledge, ideas and reflection more widely available and legitimising voices and perspectives that are currently silenced.

It is against this background and a strong commitment to the OA (Open Access) Movement, that Maynooth University Library decided to make *Silence Would Be Treason: Last Writings of Ken Saro-Wiwa* freely available on open access and also to organise events where the issues raised contribute to shaping the public discourse. Globally, there are a number of similar initiatives. In an effort to publish as part of the commons, when the author and/or co-publishers are agreeable, Daraja Press is making available for free online, the entire text of their recent books and ebooks.

The first edition of *Silence Would Be Treason: Last Writings of Ken Saro-Wiwa* (Corley, Fallon, Cox, 2013) was published by Daraja Press in 2013. The 2nd revised edition of the book was launched in 2018. It contains additional material including a preface written by Noo Saro-Wiwa, a chapter detailing PhD student Graham Kay's research on the historic links between government and the petrochemical industry, a chapter by Dr Anne O'Brien and Helen Fallon on the open access Ken Saro-Wiwa Audio Archive, and an afterword by Mark Dummett, Business and Human Rights Researcher at Amnesty International. Both books are available in print format and on open access and can be downloaded from the MU institutional repository MURAL (Maynooth University Research Archive Library).

In a letter to McCarron, dated 1st December 1993, Saro-Wiwa wrote:

Keep putting your thoughts on paper. Who knows how we can use them in future. The Ogoni story will have to be told.

(Saro-Wiwa, K. 1993)

Open Access is ensuring the story is told and is accessible to all.

References

Cochrane, L. and Lemma M.D. (2019). Reflections on open access from the Global South – Melisew Dejene Lemma. *Nokoko: Journal of the Institute of African Studies,* Carleton University, Ottawa, Ontario. https://ojs.library.carleton.ca/index.php/nokoko/article/view/2346

Collyer, F. M. (2018). Global patterns in the publishing of academic knowledge: Global North, global South. Current Sociology, 66(1), 56–73. https://doi.org/10.1177/0011392116680020

Collyer, F.M. (2018) Ibid

Corley, I., Fallon, H., Cox, L. (2013). *Silence Would be Treason: Last writings of Ken Saro-Wiwa.* Senegal: CODESRIA/Daraja Press. http://mural.maynoothuniversity.ie/8940/

Corley, I., Fallon, H., Cox, L. (2018). *Silence Would be Treason: Last writings of Ken Saro-Wiwa.* 2nd edition. Senegal: Daraja Press. http://mural.maynoothuniversity.ie/10161/

Larivière V, Haustein S, Mongeon P (2015) The Oligopoly of Academic Publishers in the Digital Era. PLoS ONE 10(6): e0127502. https://doi.org/10.1371/journal.pone.0127502

Saro-Wiwa, K. (1993). Letter of 1st December. Maynooth University Ken Saro-Wiwa Archive PP7/2

Embedding Archival Collections into the Postgraduate Curriculum at Maynooth University Library

Ciara Joyce and Roisin Berry

Introduction

The Ken Saro-Wiwa Archive is one of Maynooth University Library's most hard-working collections. Since acquiring the archive in 2011, it has been utilised as part of a wide range of outreach experiences, including seminars, exhibitions (nationally and internationally), blog posts, articles, posters, books and book chapters, poetry workshops, school visits, and guided tours. In 2019, the collection was the focus of a workshop delivered as part of the module *Peace, Religion and Diplomacy* offered on both the MA in Mediation and Conflict Intervention and the MA in International Peacebuilding, Security, and Development Practice at Maynooth University (MU). This essay describes this initiative.

Unique and Distinctive Collections as a catalyst for learning

New approaches to skill acquisition and critical thinking using primary sources are increasingly common across campus and beyond. They present, to the Library and the academic community, exciting ways to engage with students, and to the archivists, a wonderful opportunity to share their knowledge and love of archival collections to an engaged and enthusiastic audience.

The Ken Saro-Wiwa Archive

The archive includes 28 letters to Sister Majella McCarron, of the Missionary Sisters of Our Lady of the Apostles (OLA), 27 poems written by Saro-Wiwa, photographs showing the destruction of Ogoni villages, video cassettes recording visits and meetings after Saro-Wiwa's death, and articles and reviews relating to his work and to the campaign to save his life. It also includes a number of artefacts including a MOSOP cap which belonged to Ken Saro-Wiwa and a MOSOP flag.

Letters

The importance of the archive is that it captures in rich detail the last two years of Saro-Wiwa's life, and documents his transition from activist to political prisoner. The letters and poems in particular record themes such as the on-going struggle to protect the Ogoni people, growing instability in Nigeria, Saro-Wiwa's conditions during his detention, and the importance of his friendship with Sister Majella during the final chapter in his life.

A letter to Sister Majella dated 13 July 1994, provides an intimate account of Saro-Wiwa's living conditions during his imprisonment. He states:

> *'My condition is not very bad. I have an air-conditioned room to myself and the electricity has only failed once. I can write and only yesterday succeeded in smuggling my computer into this place. I can cook (though I cannot cook) for myself and from time to time, I can smuggle out letters … The only thing is that family members, lawyers and doctor are not allowed to see me.'*[1]

To the fore of Saro-Wiwa's mind, however, is always the plight of the Ogoni people. In letter after letter, he expresses his concern regarding their welfare. His commitment to his people and their protection is unwavering, even when he himself is faced with death. In a letter dated 29 October 1994, he states:

> *'My moments of depression here had more to do with the political situation in the country: worries over the Ogoni and such-like than the fact of my con-*

1 Maynooth University Library, Ken Saro-Wiwa Archive, IE/MU/PP/7/3

finement. I miss my family, of course, but…it is a fitting price to pay for the joy of others.'[2]

Throughout his detention, and during some of the darkest chapters in his life, Saro-Wiwa reached out again and again, to his supporter and friend, Sister Majella McCarron or 'Sister M' as he often referred to her in his letters. A quote from another of his letters, dated 1 October 1994, captures the importance of their friendship to him. He tells her:

> *'I long to see you back in Nigeria, helping among others, to guide the Ogoni people….You don't know what help you have been to us, and to me personally, intellectually.'*[3]

Archives Workshop

In early 2019, the Edward M. Kennedy Institute at MU approached the Special Collections and Archives department at MU Library about the possibility of using the Ken Saro-Wiwa Archive to deliver part of a module entitled Peace, Religion and Diplomacy. The Institute, established in 2011,

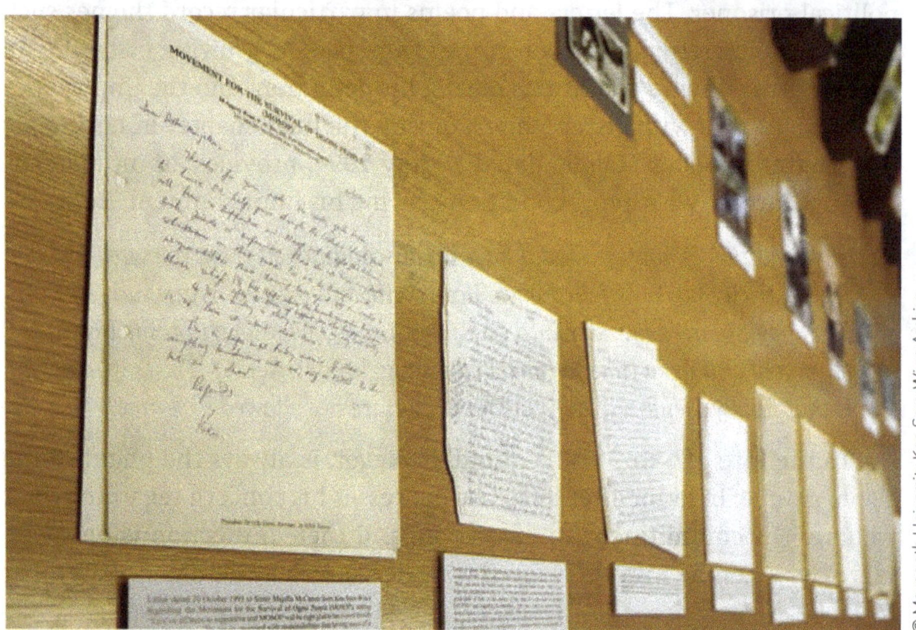

Items from Ken Saro-Wiwa Archive

2 Maynooth University Library, Ken Saro-Wiwa Archive, IE/MU/PP/7/13
3 Maynooth University Library, Ken Saro-Wiwa Archive, IE/MU/PP/7/18

aims to 'build capacity for constructive approaches to conflict at all levels in society'.[4] Through a series of lectures and workshops, the module addresses freedom of thought, conscience, religion, or belief, and raises questions regarding religion and society from the perspective of diplomacy.

The module focuses on the role of religion and human values in international relations, looking at how religion can contribute to 'moral discernment in difficult circumstances'.[5] Students are asked to look at the role of the 'courageous individual who perceives the underlying truth of situations and works with others to bring the truth to attention'.[6]

In looking for suitable examples, Philip McDonagh of the Edward M. Kennedy Institute decided the Ken Saro-Wiwa story was a fitting illustration of someone who 'separated truth from falsehood and took a stand for justice in the face of seemingly invincible social forces'.[7] Saro-Wiwa's friendship with Sister Majella and the consolation that relationship brought to Saro-Wiwa in prison was also relevant to the module.

In 2019, two workshops were offered on the theme of conscience and the individual, one looking at the case of Ken Saro-Wiwa and one at the case of Franz Jägerstätter, a World War II conscientious objector and subject of the 2019 film "A Hidden Life." Students were then asked to compare the stories of the two men.

Deputy librarian Helen Fallon and archivist Ciara Joyce delivered the Ken Saro-Wiwa workshop to participating students in the Special Collections and Archives Reading Room, in March 2019 and again in December 2019.

Methodology

The two-hour workshop began with an introduction to Ken Saro-Wiwa and the Ogoni cause. Students were given the context of Saro-Wiwa's campaign, arrest and execution and the background to how the letters he wrote to Sister Majella found their way to Maynooth.

Students then had the opportunity to look at the original letters, poems and photographs on display in the Reading Room. Archivist Ciara Joyce spoke about each item on display, reading out relevant extracts. The items were carefully chosen based on their content and the required learning outlined in the course descriptor. Students were asked to consider the complexity of the personal, political, economic and legal issues facing

4 https://www.maynoothuniversity.ie/edward-m-kennedy-institute/about, accessed July 2020
5 Personal correspondence to the author from Philp McDonagh, Edward M. Kennedy Institute, June 2020
6 Ibid
7 Ibid

Saro-Wiwa or anyone who takes a 'counter-cultural' stance. Each letter exhibited showed evidence of these issues. For the rapid dissemination of the contents of the documents and to aid the discussion, students received a handout highlighting relevant extracts from the letters.

The students undertaking this module were extremely enthusiastic, well-informed and engaged with the topic. The subsequent discussion demonstrated both their knowledge of the Saro-Wiwa story, and that they had thoroughly considered the implication of the stance he made and his personal sacrifices. Four of the students were from Nigeria and were able to offer their unique perspective on Saro-Wiwa's writing and the contents of the letters, which was interesting for both participants and library staff.

Outcomes

The workshop afforded the opportunity to critically examine a complex and multidimensional topic. The original letters and the issues they embody were used as a catalyst for the discussion, demonstrating their usefulness in the development of investigative and interpretive skills in the students and in stimulating informed group discussion.

For some of the participants, it was their first visit to the Special Collections and Archives Reading Room and it demonstrated to them the importance of preserving and consulting primary sources.

Overall, the workshop was very enjoyable and worthwhile for both students and staff and the Library received very positive feedback from the participants. Three of the seven students decided to use Saro-Wiwa for their module assignment, indicating the high level of interest in the contents of the workshop.

As an archivist working with this archive for several years, it is very rewarding to see such enthusiasm for the collection and see how directly relevant the material is to the students' course work. By presenting an active learning experience we hope it somewhat demystifies Special Collections and Archives and will increase student engagement.

The use of unique and distinctive collections by university libraries in teaching is becoming increasingly common. With the growing emphasis on critical thinking and skills development, this is likely to increase in the future. The Saro-Wiwa archive had previously been used in the teaching of an undergraduate module in the MU BA in Community Studies in 2013[8]

8 Fallon, H & Ryan, A (2014) Death Row Correspondence: Integrating the Ken Saro-Wiwa Collection into the Undergraduate Curriculum – http://mural.maynoothuniversity.ie/5550/

and in the teaching of Post-Colonial Studies. The Teresa Deevy archive, the Pearse Hutchinson archive and the Littlehales archive are just some examples of archival collections at MU Library used in both undergraduate and postgraduate classes.

Through this collaboration with the Kennedy Institute the Library played a significant role in a postgraduate programme that seeks to examine and influence the changing character of diplomacy and international relations, and the successful outcome to date indicates that it will be an ongoing collaboration.

An earlier version of this essay was published in Archives and Records Association, Ireland, Spring 2019 Newsletter

More information about the Maynooth University Ken Saro-Wiwa Archive can be found in the online Ken Saro-Wiwa library guide.

Whose story: Working towards diversity in the Maynooth University Library Collections

Hugh Murphy

Introduction

The University campus in Maynooth is home to two third level institutions – Maynooth University and St. Patrick's College Maynooth. Maynooth University was formally established as National University of Ireland Maynooth in 1997. It has its origins in St. Patrick's College, a seminary established in 1795. There are two buildings, the modern John Paul II Library and the historic Russell Library. Thus the Library in its various incarnations has a history of several centuries and this is reflected in its collection. The original, foundational collections, used by the emergent college in the early 19th century, were in many ways representative of a classic enlightenment era library, with material contributed by the early members of the college community, many of whom had been based in seminaries in France and Spain.

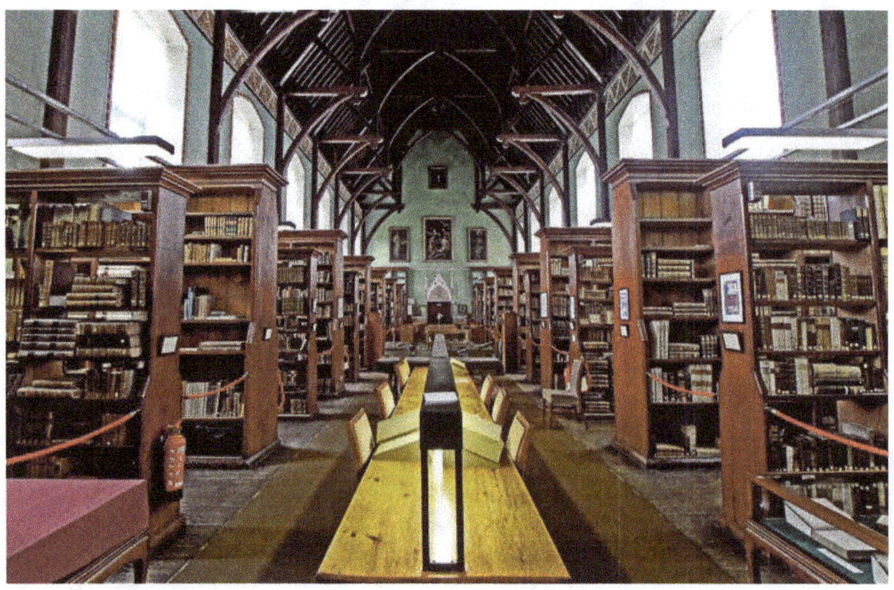

Russell Library

These were added to over the 19th and 20th centuries, often by bequest and donation from priests and alumni scattered around the globe.

St. Patrick's College opened its doors to lay students in 1968. In 1997 there was a formal split with the establishment of National University of Ireland Maynooth (NUI Maynooth), subsequently termed Maynooth University. The last decade has seen a renewed focus on unique and distinctive collections, which are housed in the purpose-built Special Collections and Archives section in the John Paul II Library. This department, which houses among other collections, the Ken Saro-Wiva Archive, is a key part of the library and represents a unique aspect of the library experience. While the library's collections are available for study and research purposes to students and staff of both Maynooth University and St. Patrick's College Maynooth, Special Collections has a broader remit, as its collections are of interest and use to a broader cohort, including academics and students from other educational institutions (both nationally and internationally) and interested parties from the general public. To this end, these collections have been actively developed with the result that the scale and diversity of our special collections is incomparable to that of heretofore, with items ranging back to pre-Christian times. Many of our special collections now form a key foundation in our support for teaching across several disciplines.

If we operate on the basis that a library is a combination of people, place, and collections, it is clear that all three have seen significant evolution in Maynooth University, as in other institutions. In recent years, the idea and the reality of library collections has been challenged in several significant ways and this merits some exploration.

Cuneiform tablet

Firstly, as noted previously, our collections are broadening in scope and type. Where we once had a print-focussed collection predicated almost exclusively on the teaching needs of the university, the modern library sees collections ranging from Mesopotamian clay tablet, through manuscript and book to digital object.

Open Access

Secondly, considerable work has been undertaken in the last decade to make scholarly literature more accessible to all. The Open Access (OA) Movement, while born in the Global North, has the potential to be transformative to institutions in the Global South, that cannot afford the punitive cost of subscription to scholarly publications, as outlined in more detail in Manji and Fallon's essay in this volume. OA can be seen as a fundamentally positive movement in scholarly publishing (and by extension in education) as it allows readers across the globe to gain access to academic scholarship, previously 'hidden' behind punitively expensive paywalls. For a researcher who is not affiliated to one of the wealthier universities, OA offers, at least in theory, an increasingly unfettered landscape of scholarly literature. In this regard it ensures that a far more diverse readership is accessible beyond the local academic community, which can only be good for inclusivity. This is evidenced by the broad reach of the open access version of *Silence Would be Treason: Last writings of Ken Saro-Wiwa*, containing the letters and poems of Saro-Wiwa to Sister Majella McCarron (OLA).[1]

The devil of course, is in the detail, and it is clear that academic publishers are typically loath to undermine their own business plans and, in some cases, exorbitant profit margins. However, even beyond this, the proposed model of OA espoused by many funders and publishers still requires payment of a sort, and this can serve as a deterrent to researchers who are either unaffiliated or who work in institutions and regions where funding is unavailable to support open publication. The challenge of the 'Eurocentric' model of Open Access for such regions is clear and have been clearly articulated by prominent research groups in Central and Latin America for example.[2] For Open Access to be truly global in its benefits, it is clearly essential that models of funding, revenue and access are agreeable to all who might benefit and can be enacted at scale from the wealthiest to the poorest region.

Digitisation

Thirdly, and perhaps less contentiously, by harnessing the opportunities afforded by digitisation, libraries can 'open' their collections, providing

1 Corley, I., Fallon, H. & Cox, L (2018) *Silence Would be Treason: Last Writings of Ken Saro-Wiwa.* Montreal: Daraja Press. http://mural.maynoothuniversity.ie/10161/

2 Statement of the first consortium assembly from Ibero-America and the Caribbean, released September 2017

free access to some of their most remarkable items to a global audience. Where once the prohibitive cost of subscription or travel might have prevented a researcher from visiting a library to view an essential collection, digitisation can enable a parity of access irrespective of geography or economics. Where Open Access is a complex series of challenges and negotiations between various stakeholders, local digitisation is library-led and arguably can be seen as the ultimate expression of the quintessential aim of a library; to have its collections engaged with, interrogated and through this to stimulate leaning and the development of further knowledge. Given the way that collections have become fragmented between repositories and indeed regions, digitisation allows for the possibility of reunification of a collection and perhaps, if necessary a degree of 'digital repatriation' where an object which was removed from a region becomes accessible again albeit virtually. Clearly this would appeal to the global Irish diaspora who might relish accessing our Gaelic manuscripts for example, but when one considers the variety of our collections, it is evident that there are other groups who, while having no connection to Maynooth or even Ireland, may have an interest in a particular collection or item. In this regard the Saro-Wiwa Archive was a clear exemplar, being of interest to both Irish and Nigerian audiences, but also students of social and economic history, social justice and activism and post-colonial literature.

Developing the collections

Fourthly and most importantly, the University has worked assiduously to consider precisely what to collect. This work came to fruition in the current collection development policy which attempted to distil our aims. Critically the policy notes the importance of the library role in representing those on the outside, or the margins of society. While it was not possible to provide a comprehensive suite of guidance, the policy endeavours to articulate the aim and the rationale of focussing on:

> "the idea of 'The Outsider' which can be said to be a multidisciplinary theme, encompassing figures from various backgrounds, who were either marginalised or viewed as existing on the fringes of contemporary society, but whose impact in areas such as literature, history, or social movements is considerable. The adoption of such thematic areas is informed by the research themes of the

University (both distinct and interdisciplinary) and the unique research heritage of both Maynooth institutions"[3]

Critically this is not simply an aspiration for its own laudable sake – there is a clear and important link to the educational aims of the university. In Maynooth this can be most visibly recognised in our unique resources, which encompass archives of less well-known playwrights, poets, marginalised historical figures and, importantly, in the collections of human rights figures such as Saro-Wiwa and Fr Dennis Faul.[4] In recent years, the Library has furthered its focus on human rights issues, acquiring collections relating to the Northern Ireland Troubles. Interestingly one of the poems written by Saro-Wiwa for Sr. Majella McCarron focuses on the link between Ogoni and Fermanagh (McCarron's birthplace in Northern Ireland), noting the shared agony and hunger for justice and peace.[5]

The Troubles collection

The key aim of the Library has been to collect – in as representative and impartial a way as possible – material showing not only the views of Catholic and Protestant, but of marginalised groups such as the British and Irish Communist Association.

Our collections have evolved over many years. The question of diversity in all its aspects, including collections, has been an area of more recent concern. Much of our modern collection is, by intent, reflective of the current teaching and research needs of the academy. In this, it can be said to

3 MU Library Collections Development Policy p. 6

4 Monsignor Denis Faul (1932-2006) was a Maynooth educated priest who became actively involved in the Northern Ireland Civil Rights Movement and the hunger strikes of 1981.

5 Maynooth University Ken Saro-Wiwa Archive 20/6/95

reflect the academic views of colleagues who order items to support teaching and research. It is challenging, but important to acknowledge however that these collections will therefore also reflect any lacuna in knowledge and unconscious biases of all involved in developing it. This is not simply a consequence of a relatively homogenous staff cohort – it is also reflective of inequities in publishing and access. Such challenges for librarians, have been catalysed in recent years as part of a broader move to consider questions of diversity, white privilege, inclusion, and the important issue of decolonising of curricula and collections in Universities in the Western World. While this has figured most prominently in regions with clear legacies of empire building, it is essential that Irish institutions reflect on the diversity of their collections and, critically, the way these collections were established.

For Maynooth, this requires us to look in particular at the role of the original educational establishment, St. Patricks College, which trained priests, many of whom partook in missionary activity around the world. Several of the more remarkable items in our collection were acquired as the result of these distant benefactors, including our stunning collection of Mesopotamian clay tablets which date back to c3000 BC. These were given to the College by an Irish army chaplain and alumnus during the First World War. The fact that Ireland was, in many ways the first colony of the British Empire, does not obviate us from our responsibility to recognise our role in empire building and certainly should not preclude library staff from trying to reconcile the tension of an inheritance of wonderful collections with (in some cases), the potential legacy of oppression which brought them to us.

In the recent past the Library has used such collection in the curation of exhibitions on topics relating to the middle and far east which might strike someone unfamiliar with both college and collections as incongruous given the origins of the college. It is essential when showcasing these collections to acknowledge their origins and indeed to strive while acknowledging, to avoid over-writing the culture and the provenance which brought them into being. With an increasingly diverse demographic of users, including international and students coming in via our 'university of sanctuary' status, outlined in Cliona Murphy's essay in this volume, there has never been a greater need for us to engage in considered reflection on our collections and how we present them.

Libraries have long aspired to be seen as 'neutral' spaces, devoid of observance to any espoused political or religious viewpoint. This does not preclude us from thinking critically on the morality of who our collections represent and how they came to be. While we rightly extoll the merits of an archive such as that of Ken Saro-Wiva, we must also acknowledge this legacy. Indeed, even the act of recognising the legacy is important, as it provides us with a foundation towards a more diverse collection which represents the breadth and variety of the world which the University seeks to engage with, learn from and contribute to.

The University as a Place of Sanctuary: Advocating for Inclusion and Equality for Refugees and Migrants

Clíodhna Murphy

Introduction

For Kurdish novelist and journalist Behrouz Boochani, writing is an 'act of resistance'.[1] In 2013, he fled persecution in Iran due to cultural activities but on arriving to the Australian territory of Christmas Island, he was exiled to an offshore immigration detention on Manus Island, where he remained for six years under brutal conditions. The novel he wrote there, sent out to a journalist via WhatsApp messages, was awarded one of Australia's most prestigious literary prizes.[2] Accepting the prize from Manus Island, Boochani described it as 'a victory for humanity'; a recognition of his art even as the system sought to obliterate the individual identities of the Manus Island detainees. He was formally recognised as a refugee in New Zealand in 2020.

As much as Boochani's journey and work has parallels with that of Ken Saro-Wiwa, it prompts us to think about the treatment of those seeking sanctuary in our own country, including within public institutions such as universities. Although applicants for international protection in Ireland (asylum seekers) are not subjected to the extreme forms of degradation outlined by Boochani in his writing, they are systematically excluded from many aspects of Irish society. They are disqualified from the mainstream social welfare system and are accommodated in 'direct provision' centres (which provide accommodation and food), receiving an allowance of €38.10 per week. While many human rights deficiencies have been identified through multiple studies of the system[3] (which the current Govern-

1 Behrouz Boochani, "Writing is an act of resistance" (TedXSydney, 30 July 2019).

2 Behrouz Boochani (translated by Omid Tofighian), *No Friend But the Mountains* (Picador, 2018).

3 For example, Claire Breen, "The Policy of Direct Provision in Ireland: A Violation of Asylum Seekers' Right to an Adequate Standard of Housing" (2008) 20(4) International Journal of Refugee Law 611-636; Liam Thornton, "The Rights of Others: Asylum Seekers and Direct Provision in Ireland" (2014) 3(2) *Irish Community Development Law Journal* 22-42; "Direct Division Children's views and experiences of living in Direct Provision: A report by the Ombudsman for Children's Office 2020" (Ombudsman for Children's Office, 2020). 4 Irish Refugee Council, "Powerless: Experiences of Direct Provision During the Covid-19 Pandemic August 2020" (IRC, 2020), p. 14.

ment has committed to ending), the situation has been particularly acute during the COVID-19 pandemic. One direct provision resident explained the situation as follows in response to an Irish Refugee Council survey:

> "We live in a much crowded place, we have to share rooms (minimum three people) and toilets, we have a small shop downstairs that is too small, in the corridors there's like a thousand doors you have to open before reaching your room. There's no way you can stay in the room without going out, the cooking stations are far away from the rooms and you have to go outside to get to the toilet. It's very hard to lock kids aged (3-11) in these small rooms."[4]

Another stated: "We are powerless, just sitting ducks waiting to die."[5]

Wider exclusions, outside of the accommodation system, also exist. Until 2017, asylum seekers were not permitted to work while awaiting the outcome of their application (even if this took many years). Despite now being allowed to work after 9 months, in the ongoing public health crisis, those who lost their jobs were excluded from the 'pandemic unemployment payment', until outbreaks in factories and direct provision centres forced a change in policy. Asylum seekers remain ineligible for Irish drivers' licences. In the higher education context, while those with refugee status may access free higher university education and student grants, asylum seekers are subject to international fees and are not eligible for the main state student support grant (although there are some exceptions for those who can access a specialised support scheme).

The Sanctuary Movement

It is in this global context of a hostile environment for asylum seekers, refugees and migrants that the City of Sanctuary and University of Sanctuary movements grew up in the UK. City of Sanctuary UK was founded in 2005 as a network of community organisations and is a mainstream grassroots movement working towards the vision that the nations of the UK "will be welcoming places of safety for all and proud to offer sanctuary to people fleeing violence and persecution".[6] Ireland's Places of Sanctuary movement is closely affiliated to City of Sanctuary UK. Universities and Colleges of Sanctuary Ireland (UoSI) emerged as an initiative to encourage

4 Irish Refugee Council, "Powerless: Experiences of Direct Provision During the Covid-19 Pandemic August 2020" (IRC, 2020), p. 14.
5 Irish Refugee Council, p.14.
6 Information available at https://cityofsanctuary.org/about/ (last accessed 10 August 2020).

and celebrate the good practice of universities, colleges and institutes welcoming refugees, asylum seekers and other migrants into their university communities and to foster a culture of welcome and inclusion for all those seeking sanctuary. The aim of the network is to spread this culture of welcome across the institutions of higher education all over the island.[7]

Universities and colleges can apply to be designated as sanctuary institutions and almost all the Irish universities now have this status. The sanctuary movement is based on broad principles of inclusion and universities of sanctuary are free to develop their own activities. However, they are required to demonstrate how their institutions follow three key principles:

- Learn about what it means to be seeking sanctuary in general, and in higher education institutions in particular.
- Take positive action to establish a sustainable culture of welcome.
- Share what they have learned and demonstrate good practice with other education institutions, the local community and beyond.

In practice, because of the aforementioned structural barriers to accessing university education, the provision of scholarships by individual institutions has been a central focus of Irish universities of sanctuary to date. Maynooth University is starting with three scholarships for the academic year 2020/21. For this pilot year, it is planned that one of these scholarships will go to a postgraduate student, and two to undergraduates (one in the field of Science and Engineering; and one in Arts, Humanities or Social Sciences). The scholarships are funded by the University and cover the students' fees, laptop, transport costs, and subsidised food, as well as a range of academic and other student supports through the Maynooth Access Programme. The spectre of borders looms large over the process: in a number of Irish universities, sanctuary students have been notified of an intention to deport and some sanctuary students have had to abandon their studies because they have been directed to move accommodation centres.

What is a university of sanctuary? The Maynooth experience

Maynooth University began the groundwork to become a university of sanctuary in 2018 and received the designation in 2020. A number of key aims were pursued during this initial period: to establish a steering committee with representatives from across the university (including students), NGOs

7 Information available at https://ireland.cityofsanctuary.org/universities-and-colleges-of-sanctuary.

and members of the migrant community; to record the type of work happening in this space at Maynooth; to set up a scholarship scheme; and to consolidate links with local direct provision centres. A university-wide call for information demonstrated that extensive sanctuary-related teaching, research and volunteer activities is already taking place in the University. The topic of migration is deeply embedded in the curriculum: it is taught in 36 separate lecture courses and is the central focus of 15 modules.

The pilot scheme of sanctuary scholarships, open to asylum seekers and refugees without access to state support for their studies, revealed an appetite for university education among this group and there were many more suitably qualified candidates for the scholarships than scholarships available (itself an important finding for the project going forward). The process has also shown us the complexity of the issue of access to and progression through third-level education for asylum seekers in Ireland. Participants at the launch of the scheme identified significant barriers to pursuing university education, including childcare, mental health issues, and accessing expensive books. The scholarship application and selection process threw up further problems relating to the accessibility of direct provision centres in places such as rural Kerry and Leitrim, as well as obstacles to asylum seekers applying to part-time courses and problems with registering for courses due to documentation issues. More broadly, the COVID19 landscape of remote teaching has raised questions about digital disadvantage and difficulties in accessing online taught material.

In terms of immediate future plans, we hope to conduct a participatory action research-based local needs assessment with the county Kildare direct provision centres to establish how Maynooth University might support access to third-level education and play a more general role in community integration. We will also conduct a thorough analysis of potential administrative or internal systemic barriers to accessing MU courses. The scope of the project will be kept under review: for example, one of the NGO representatives on the sanctuary committee has suggested that it should be expanded to specifically encompass undocumented migrants who face significant obstacles to accessing university education.

Difficult questions

As a University community, through the Sanctuary process, we have agreed to the goal of inclusion and welcome for asylum seekers, refugees and migrants within our institution. However, the experience so far has shown that the devil is in the detail: how do we actually achieve these goals? How do we adequately support students in relation to the complex issues of access to childcare and mental health supports? Do the scholarships serve to relieve pressure on the Government to radically reform its policy on university access, thereby becoming counter-productive? Should the national Universities and Colleges of Sanctuary network advocate for a sectoral commitment to charging this relatively small cohort of students EU fees (rather than the much higher international fees)? How does the sanctuary movement interact with a commitment to race equity and broader issues of inclusion and belonging on campus?

As quickly realised by the sanctuary committee at Maynooth, there must be an ongoing process of reflection on what it means to be a university of sanctuary.

From a personal perspective as an academic working in the field of migration law and human rights, it will be interesting to see how the goals of inclusion and welcome fit with the more politically-charged aims of justice and equality and to think about the appropriate role for academic institutions in pursuing these objectives. For now, the University of Sanctuary process has opened up a space to explore these conversations in a more structured way across the university and in dialogue with students and the community.

Nothing About Us Without Us

Nigerian and Irish Women Working Together

Veronica Akinborewa, Camilla Fitzsimons and Philomena Obasi

This chapter is probably quite different to what comes before and after it. It is a conversation between three adult educators; Veronica who is a Nigerian woman of colour, Camilla who is a white Irish woman and Philomena who is a woman of colour, also from Nigeria. We have known each other for around 5 years, firstly through a student-lecturer relationship and more recently through work we have undertaken that involves the design and delivery of workshops on 'culture', 'interculturalism' and 'racism'. Our approach contrasts with how many of you might more typically read 'about' these topics. We lean on an autoethnographic approach where our writing seeks to describe our experiences and draw out analyses from these. This method is thus both a process and a product (Ellis, 2011). Our first-hand account explores relationships that exist amidst the intersections of race, gender and institutional positions. Our capacity to overcome these barriers and successfully work together is determined by the quality of our relationship. Nothing is taboo and we have often discussed the joys and challenges of working together.

Here we share a flavour of just one of these discussions with you. It was recorded during the Covid19 lockdown of 2020 when each of us were confined to our homes and connecting via Zoom.

On identity....

Camilla: I might start, or anybody else can, I don't mind.

Veronica: No, you go ahead, take the lead Camilla please.

Camilla: That's part of the story, isn't it, me taking the lead? Sometimes I worry about that, but we can talk about that later. So, I'm Camilla, I'm Irish. I grew up in an environment where I never questioned what it meant to be Irish, there were very strong cultural markers; being Irish in the 1970s meant being white, eating potatoes, going to mass, that sort of thing. I also grew up during 'the troubles' in Northern Ireland. There was a lot of bombings as I remember it, lots of people getting killed. That kind of fed into my Irish identity. Who's next? Philomena, do you want to go next?

Philomena: Yes okay, I can go next. My name is Philomena Ilobekeme Obasi, I'm from Edo State, Nigeria. I had my son in Ireland in July 2001 and my daughter in America. My children had their primary education in Nigeria. We relocated back to Ireland in 2014 so they could continue their secondary education. I had heard about Ireland growing up. I had always loved Trinity College because all the missionaries (priests and nuns) that came to Nigeria from Ireland always talked about it.

Veronica: Okay me. Veronica Adeyinka Akinborewa, originally from Osun State, in western Nigeria. I came to Ireland in 2002 with my first son. I had my second son and my daughter here. When I arrived in Ireland, we were not allowed to work until we got our residence permits. We just stayed at home and were not sure of what the future held for us. So, I volunteered with a charity organisation and got involved in some integration projects in my local community.

When I got my residence permit, I felt it was time to try and get a job and I assumed that I was going to get a 'good job,' if there is anything like that. I already had a first degree in Africa before coming to Ireland. It was not long before I realized that I was building castles in the air! As a matter of fact, it seemed as though the 'proper' jobs were meant for our Irish counterparts while the menial jobs were meant for us, 'the others.'

So, what did I do? I reverted to my baptismal name 'Veronica' so the recruiters might at least give my CV attention and, lo and behold, the interviews started coming in. I did this because I read an article about a young Nigerian-British lady who had changed her name to a more English name so she could get a job. I'm not proud of saying this, because I really do love my traditional name, my native name. But I thought at that time, because I was desperate to move on, I was desperate to…get into a better position. I just did not have an option at that time.

Camilla: And you were not imagining it. There was an experiment by McGinnity *et al* (2008). They made up fictitious CVs with different names; either African, Asian or European but with similar qualifications then sent out. The candidates with 'Irish names' were twice as likely to get called for interviews.

The politicising nature of our lived experiences

Philomena: I had my degrees from Nigeria, a Bachelor in Education, Masters in Public Administration and a Masters in Business Administration. Although teaching was my principal profession, I ended up working in the finance sector and oil and gas.

When we relocated to Ireland in 2014, my children were initially going to school and I was mostly at home. I said to myself 'okay, I need to integrate, I need to meet Irish people, especially women.' So, I volunteered with the Parent-Teacher Association and when my children were at school, I attended some adult education classes, some of these were at level 3 and I got to know other women doing these basic education classes. From my interactions with these Irish women, I got a sense of the reasons they were availing of the programme. Most were dropping their children into the centre's creche, then coming into these classes. Many had stopped working when they had kids or had left school very early. I saw a lot of similarities between these Irish women and many Nigerian women. I thought these inequalities were part of Nigeria's culture and didn't expect to see the same sort of thing in Ireland.

Camilla: You said you worked in oil and gas, tell us a bit more about that? I'm just conscious because there is a cross-over to the work of Ken Saro-Wiwa. I've heard you give a talk about his life in the past. Were you working for, or against, the big corporations?

Philomena: I was working as a buyer for a French oil and gas company involved in offshore drilling. I didn't know much about what was going on before then, but I discovered they were making huge profits. There were various indigenous companies and factories in Nigeria that were producing most of the goods and services that were being imported from France, Germany, the United States, etc. This made me question the need for what seemed as much as an 80% importation when we could easily source most of the products in Nigeria.

Camilla: Sounds like it's quite a politicising experience?

Philomena: Yes, it was. The indigenous people of the oil producing areas were/are living with a lot of poverty – no good pliable roads, clean drinkable water, schools, electricity, and local amenities. Before the emergence of oil companies, the people were predominately fishermen and farmers as they live in riverine areas. The oil spillage, due to the negligence of these companies caused massive environmental damages, this is what Ken Saro-Wiwa and his group were often protesting against.

Camilla: We must talk more about this again. I guess my interests are mostly about 'racialisation' and racism. I used to do workshops about 'culture' and 'interculturalism', but I began to feel quite uncomfortable about this. I got tired of being in rooms with mostly white people from the dominant culture, 'othering' people. I couldn't stop thinking of the expression 'nothing about us without us.' I started reading contributions by bell hooks, she's African American. She talks a lot about the intersections of sexism and racism and says 'white people' have been socialised to see black women as 'less than', people who should be cleaners or carers; what she calls 'symbolic mammies' (hooks, 2004, p. 99).

Philomena: When I told people I might go back into teaching, there was some resistence from both my Nigerian and Irish counterparts. Some people said things like 'Oh you want to teach, a black person teaching in Ireland?' or 'Why don't you go to healthcare? you want to take our jobs from us?'

Camilla: Sounds like hooks is onto something.

Philomena: There's also a website by The Migrant Project that talks about ways in which racism affects Nigerian people's ability to get work.[1]

Camilla: Interesting. I started to think about my own whiteness in a much more deliberate way. I like the ideas of Frances Kendall (2013). She's also white. She talks about the invisibility of the privilege I grew up with and says white people have to think more about this. So, I began to write about what it means to be white and from the dominant culture (Fitzsimons, 2019). For me, this is the most authentic contribution I can make.

And I know that's not necessarily your interests, that's another thing I worry about, that the black people in education are mostly asked to talk about their culture and ethnicity and not about anything else. I would love to see much more integration and we don't seem to have quite gotten there yet.

Philomena: We desperately need black and migrant teachers as educators!

Veronica: Growing up in Africa, I always had a passion for working with communities who have been marginalised. Here in Ireland, I found myself working at different levels, firstly with people living in what are sometimes described as 'poorer areas' and later with men recovering from addiction. So, I decided to do a degree in addiction studies, to better understand their

1 https://www.themigrantproject.org/nigeria/life-in-europe/. 24 July 2020.

issues. Many of them were early school leavers, some of whom had great potential. I returned to college again and did a postgrad in teaching.

I'm also interested in culture and interculturalism. We were one of the first black families in my community. I did some voluntary work with a community group to put an integration programme together. Along the way they discovered I am a choir member in my local church and my husband is the keyboardist. They invited us to sing to older people every Tuesday evening. We would go there, we would sing, we would dance, they would sing their Irish songs and we sing our Nigerian music. And as time went on, we brought other Nigerian families that lived nearby. I also did some voluntary work with Rotimi Adebari, the first black mayor in Ireland, who was also interested in integration and was lecturing in UCD. More recently, it has been a great pleasure delivering collaborative workshops alongside Camilla and Philomena.

On working together…

Philomena: I have always wondered what informed your decision to ask Veronica and me to collaborate with you to create the degree module *Exploring Ethnic Ireland* back in 2018. I know you have told us before.

Camilla: I got an email from a colleague asking me to deliver this module to an evening adult learning group. I had a strong, almost allergic reaction to the request. Immediately I thought that it would be totally wrong for me to deliver this module alone. I couldn't deliver a module called 'Exploring Ethnic Ireland' as a white person from the dominant culture! I could do bits of it, the bits about being white, but I know nothing about being a person of a minority ethnicity in Ireland.

Sometimes I wonder if a few years ago, I might have given it a shot, but I have matured enough in my ideas to know I needed help. You were both past students on an Initial Teacher Education programme I coordinated, and I knew you were skilled adult educators. I wasn't just choosing a 'random black person' [am I allowed to say that?]. But I was conscious of what Kehinde Andrews (2018) says – that white people must use our privilege to create opportunities for people who are racialised, the people who so often get overlooked. Its back to that earlier conversation, about what happened when CVs with different names were sent out. I wanted to be part of expanding the staff of the university beyond white-middle-class people. This module needed 'white' and 'black' people working together, migrants and non-migrants. I mean it's problematic language. I know that, but I am using it anyway. It's shifting all the time and it can be hard to keep up. So, that's what informed the decision, I just kind of felt 'why on earth am I being asked to do this?'

What was it like to be asked?

Veronica: I panicked initially. Back home you refer to your lecturers as a superior. When Dr Camilla invited us to come and teach, this was very much going through my mind. I wasn't feeling confident calling her by her first name, but I was also thinking 'who would listen to a black teacher?' I have always had that at the back of my mind.

Philomena: I felt so honoured. I felt so privileged amongst a few. And I felt really happy, that this is a sign that there is change. That they will be hearing from us directly, speaking about the things we know naturally.

Camilla: I am also working with you guys on other projects, Philomena on some research and Veronica you are now part of the staff team on that same Initial Teacher Education programme you came through as a student. How do the students react when they meet you? Is there ever like 'oh my god she's Nigerian?'

Veronica: Seriously, I have not had any negative reaction from staff or students in any of the centres. I remember visiting a centre where I was once a student. The manager gave me a warm welcome and she introduced me to the other staff members. It was a wonderful feeling and it felt like a sign that we are taking some small steps in the right direction.

On racism.......

Camilla: But my sense from working with you both is that it's not always like that? This is so topical now because of the death of George Floyd and the explosion of the Black Lives Matter movement. We did a class on the Irish Black Lives Matter movement in 2018. I remember talking to the all-white degree students about how our experiences are so different. We don't have to worry about racial slurs when we walk down the street.

Philomena: This issue about race is so strong in this part of the globe. Most people in Nigeria are not aware of it. It was only when my children came to Ireland they found out about racism. They couldn't understand why their classmates asked questions like, 'Do you eat a lot of bananas in Nigeria?' They would reply, 'We don't eat banana only; we eat varieties of food.' 'Do you have traffic electricity/roads in Nigeria?' My children did not know it was racist. I would just say, 'Maybe they don't really know; they have not been there.' Most people I have met are good people, but some are racist. It really doesn't bother me. I guess it must be difficult to all of a sudden cope with an influx of migrants to once enclosed communities.

Veronica: It's sometimes subtle but yes, there is racism in Ireland. Just the other day, I was filling out a form and one of the questions on the list was about my nationality. I was glad to write that I am Irish, but I was asked for my 'country of birth', which of course is Nigeria and then I thought to myself, why do they need that information?

Camilla: Good question Veronica, or should I call you Adeyinka? Sometimes I feel that I am more powerful in our conversations. I worry about that. Maybe it's because I am the 'doctor' rather than the white person, maybe it's not real, maybe it's in my head. I do wonder if I need to be doing more to make sure I don't have to be the boss.

Veronica: I think it is a good thing that you are always so aware. That explains why it has been easy for us to work together, even though we come from different cultural backgrounds.

Camilla: And… Philomena, do you know who makes me think of you and your interest in diversity in sport? There's this young footballer who plays for Ireland and Southampton [in the English league]. Michael Obafemi, is that a Nigerian name? Apparently, he could have played for England, Ireland or Nigeria but he picked Ireland. He has a senior cap.

Philomena: Yes. He is Irish with Nigerian parents. With time, there will be a beautiful blended culture in Ireland. We look forward to that.

Bibliography

Andrews, K. (2018). *Back to Black: Retelling Black Radicalism for the 21st Century.* London: Zed books.

Ellis, C., Adams, T.E & Bochner, A. P. (2011) Autoethnography: An Overview. Qualitative social research. Volume 12, No. 1, Art. 10

Fitzsimons, C. (2019). Working as a white adult educator, using our own life stories to explore asymmetries of power and privilege. *Studies in the Education of Adults,* 88-100 doi:https://doi.org/10.1080/02660830.2019.1587876

hooks, b. (2004). *Teaching Critical Thinking: Practical Wisdom*. New York and London: Routledge.

Kendall, F. (2013). *Understanding White Privilege; Creating pathways to authentic relationships across race*. 2nd Edition. New York: Routledge.

McGinnity, M., Nelson, J., Lunn, P., & Quinn, E. (2008). *Discrimination in Recruitment, Evidence from a Field Experiment*. Dublin: The Equality Authority.

Promoting a Culture of Equality

Diversity Training at Maynooth University Library

Helen Fallon, Laura Connaughton and Edel Cosgrave

This essay discusses and evaluates a workshop on supporting integration and diversity delivered to Maynooth University (MU) Library staff in February 2020.

Background/Context

Participation in Irish higher education is changing. The number of international students coming to Ireland increased by 45% between 2013 and 2017, with the number of residence permits issued increasing from 9,300 to 13,500 over the same period. These students were primarily from the United States, China, Saudi Arabia, Malaysia and Canada (Groarke, 2019). There has been a concerted effort to attract international students to Ireland, due in part to the substantial fees paid by these students.

In addition to international students, many people have moved to Ireland to work, and/or to seek asylum. As well as an increase in migrant-students, there is also greater diversity in terms of Irish ethnicity which has contributed to a growing number of Black, Asian and Ethnic Minority (BAME) students in Irish higher education. Figures from 2017/2018 indicate that while 85.8% of students identified as Irish, the remaining 14.2% identified as other ethnicities as illustrated in the chart below. (Higher Education Authority, 2018)

Ethnic Group of Respondents, 2017/2018	Universities	Institutes of technology	All institutions
Irish	85.8%	83.2%	84.9%
Irish traveller	0.2%	0.2%	0.2%
Any other white background	6.9%	8.7%	7.5%
African	1.7%	3.1%	2.2%
Any other Black background	0.2%	0.2%	0.2%
Chinese	1.3%	0.5%	1.0%
Any other Asian background	2.2%	2.3%	2.2%
Other	1.7%	1.9%	1.8%

In this more diverse environment, cultural competency for library staff is vital. Cultural competency is defined as "a developmental process that evolves over an extended period and refers to an ability to understand the needs of diverse populations and to interact effectively with people from different cultures." (Mestre, 2010, p. 479)

Goal 5 of the Maynooth University Strategic Plan 2018-2022 has the following objective:

> To create an environment that promotes equality, diversity, inclusion and inter-culturalism
>
> (MU Strategic Plan 2018-2022, p. 45) [1]

The Library Strategic Plan 2020-2023 has *Equality, Diversity, Inclusion and Inter-culturalism* as one of its six strategic areas.

The training programme described in this essay is part of a process of increasing awareness and skills and encouraging an ongoing dialogue on inclusion. Other elements of training in this area were a number of one-hour sessions in what is known as our "Library Outside In" series. This included:

- a briefing from the MU Access Office, on a range of initiatives to bring more students into the University via a variety of targeted access programmes with schools in disadvantaged areas, prisons, traveller groups and other underrepresented groups in our society
- a briefing by the Equality Officer on the Athena Swan Programme which aims to bring more women into senior positions in higher education
- a briefing by the new MU Vice-President on her role within the University.

Most frontline staff undertook autism awareness training in 2019, and the Library piloted an online disability training course for the University in 2013 (Mellon, Cullen & Fallon, 2013).

It is against this background the programme was designed and delivered.

1 Maynooth University Strategic Plan 2018-2022 https://www.maynoothuniversity.ie/sites/default/files/assets/document/Maynooth_University_Strategic_Plan_2018-22_0.pdf (accessed 17 August 2020)

The Programme

The one-day programme was delivered in two cohorts with approximately 25 staff attending on each day. The programme was designed by the Department of Adult and Community Education at Maynooth University, in consultation with the Library. There were three facilitators. One is a full-time lecturer in the Department of Adult and Community Education in MU. Two are part-time lecturers in Adult and Community Education and are of Nigerian ethnicity.

Pre-Workshop

Before the workshop, participants received the following information, via e-mail, on the course:

As a group, participants will have an opportunity to:

- *Explore their own cultural and ethnic identity.*
- *Consider the implications of working in intercultural settings.*
- *Learn about theories of interculturalism.*
- *Understand the origins and impacts of racism.*
- *Reflect on their own way values, beliefs and theories of intercultural work and of diversity.*
- *Consider ways in which they might create conditions for supporting and celebrating diversity*

The workshops will be highly participative and experiential. Handouts and resource materials for developing ideas that might inform people's practice will be provided.

The Workshop

The workshop ran from 10 a.m. to 4 p.m. with one hour for lunch.

Morning Session

The first exercise involved staff introducing themselves and saying something about their name. Group work followed with lively discussion and flip chart presentations. Topics discussed included: cross-cultural conversations, definitions of culture (surface versus deep culture), the concept of white privilege, the difference between race and ethnicity, the language of diversity, how to have conversations around race and ethnic diversity without feeling awkward and how to deal with difficult intercultural encounters in situations where there may be a danger of being perceived as racist. The issue of how to encourage more diversity among library staff without using quotas was also discussed. Questions which arose in the discussions across the two days included: What are the unconscious biases BAME people experience in Ireland today? How are they manifested? Is asking someone to slow down or repeat what they have said potentially offensive if they are from a different culture? How do you ask someone about their background without being interrogative or making assumptions? How does the Library address the challenge of getting participants to integrate and mix more during information literacy session?

Afternoon

The participants had the opportunity to read and consider *Creating Intercultural Learning Environments: Guidelines for Staff within Higher Education Institutions* (HE4u2 (2017). This was followed by group work on how the Library can incorporate diversity into what we do. Suggestions included signage in languages beyond Irish and English, a more diverse library website and social media presence, revisiting our collection policies to ensure our books and archives reflects a variety of cultures, and aligning our outreach events with different cultural events. While we celebrate Africa Day, it was felt that this could be broadened to include Chinese New Year, Thanksgiving and other significant events from different traditions. The need to consider diversity in our teaching and training was also identified. The suggestion was made that names and examples in PowerPoint presentations should reflect our diverse users and our orientation programme for new students should be inclusive and representative of different ethnic backgrounds. Similarly, artwork throughout the building should be representative of different cultures. Culturally significant collections could be highlighted and might be used to encourage more diverse audiences at library events that are open to the public. On-going staff training in cultural diversity was also identified as a need and the suggestion that the Library have a diversity committee was made by one group.

Evaluations

An evaluation form, used by the course presenters, was distributed at the close of each day. There were 44 responses in total and feedback was very positive.

Participants rated the following statements on a "strongly agree" to "strongly disagree" scale.

1. Facilitators presented materials in a clear and organised manner
2. Opportunities to reflect on and discuss the material covered, if appropriate
3. The manner of teaching and facilitation
4. Your contribution to the learning of the group

Workshop Feedback

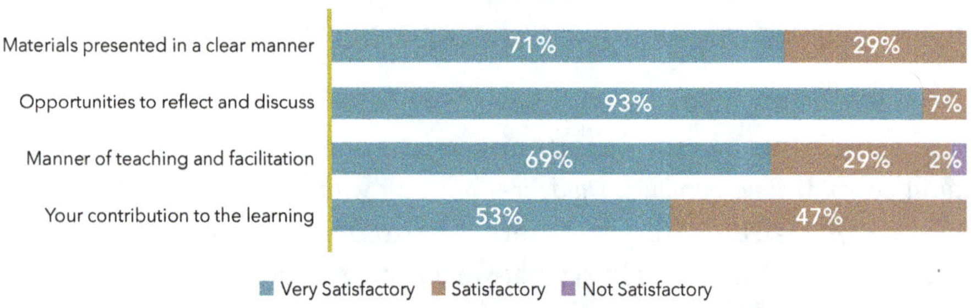

This was followed by open questions:

What did you like most about this session?

Almost all respondents provided a response to this question (44 of 45). A significant number (39%) identified the opportunity for group discussions as the aspect most liked. Several found the session informative and some mentioned learning in relation to the use of language in particular. It was also felt that the relaxed and informal atmosphere provided helped to facilitate participation and open discussion.

Liked Most

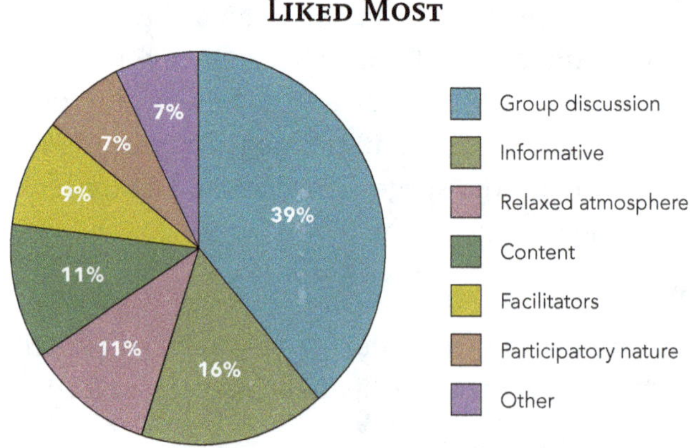

What did you like least about this session?

Some participants (11%) felt that the workshop was too long. Others felt that some topics were rushed while too much time was given to others. Some remarked (9%) that while discussion was interesting, they felt it lacked focus and direction at times.

Liked Least

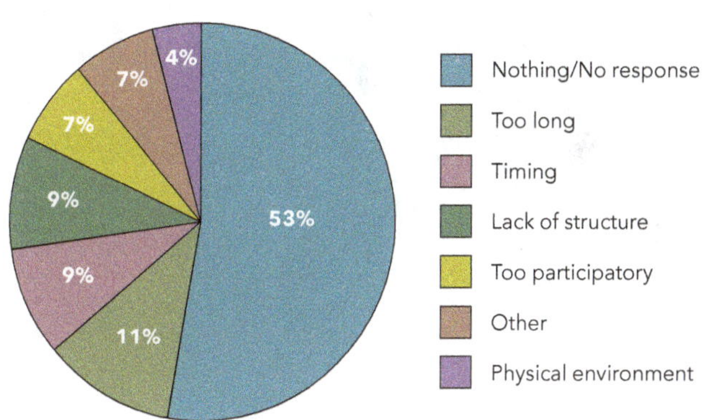

How could this workshop be improved?

Some respondents (15%) suggested that it might be useful to have representation and hear from other minority groups on campus, such as travellers, people with disabilities and Asian students. Some (11%) felt that the session could be more structured and focused.

Suggestions for Improvement

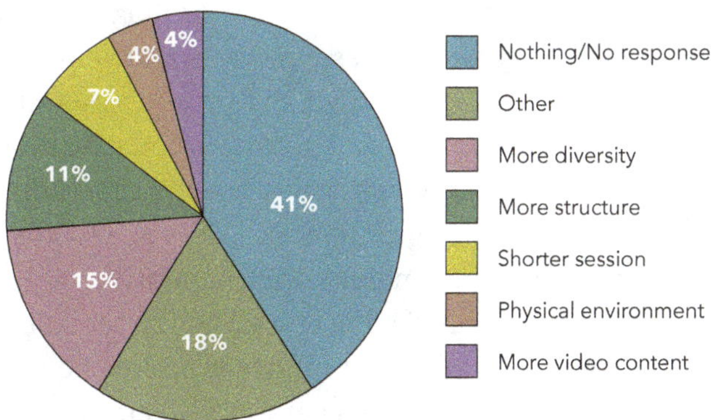

Any Other Comments?

A large number of participants took the opportunity to express their thanks for the workshop and several mentioned the lovely ethnic lunch!

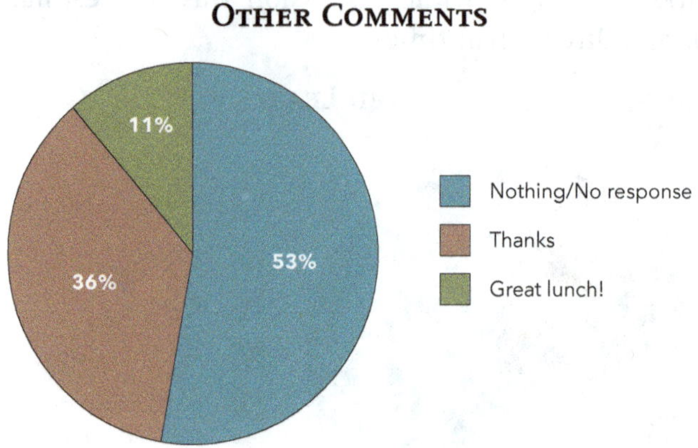

Discussion

Running a training staff for all library staff presents challenges. An obvious challenge is the need for services to continue, which means that not all staff can in fact attend. Less obvious is the challenge of providing one training programme to quite a diverse group. Frequently, individuals or groups in the Library attend specific training which relates to their role, for example most frontline staff attended autism awareness training in 2019.

The attendees have different work and life experiences. There is a large variety of library roles represented in the group. Some participants are involved in information literacy training. These people have experience of different teaching and training methods, although few would have significant knowledge or experience of the work of Brazilian adult educator Paulo Freire. This methodology places emphasis and value on the experience, knowledge and learning needs of the group, as articulated by the individual members of the group, through discussion. The programme was built around that framework, which relies heavily on group work. Some felt stimulated and engaged by group work, while others found this challenging and possibly confusing and/or boring/repetitive.

Suggestions for further improving the programme, from the authors of this paper, include developing the information on the pre-course flier and emphasising the need to read this before the course. Between the first and second offering of the programme, the flier was recirculated. One of the facilitators briefly explained the methodology to the group on the second day, following feedback from the first day, and this worked well. We would also suggest twenty as the maximum number for the one-day programme. This was in fact the number initially suggested by the presenters.

While the afternoon discussion on diversity and the Library yielded a lot of useful information, it would have been useful to integrate the objectives relating to diversity in the Library Strategic Plan 2020-2023, more into the discussion. However, this document, while drafted, had not been ratified by University Executive at that time.

Overall feedback was very positive with comments such as:

- It opened my mind
- I liked the relaxed informality and honesty of discussion and presentation
- Opportunity to participate and discuss topics was great/ Loved the participatory nature. I'm usually loathe to contribute but was comfortable to do so today due to the way it was run
- Learning from each other/interactivity made people engage more and develop more dialogue
- Use of phrases/language very helpful/made me think about how I use language
- Gave us food for thought about hidden barriers

Conclusion

Creating a racially diverse, inclusive library environment involves having conversations, which can be difficult or uncomfortable, about race and culture. It also involves looking critically at our practice and policies. We need to identify what works and what needs to change and base decisions on evidence and analysis. Both robust data and student narratives have been identified as key to progress (UUK, NUS, 2019). There is no one model for success and initiatives need to be context specific and recognise nuances of different groups. Enhancing knowledge and skills of staff, alongside improving institutional processes is vital to the process.

While this case study is MU Library specific, the model could be considered by other groups including CONUL Training and Development, the LAI Continuing Professional Development (CPD) group and other library related bodies. Diversity and inclusiveness is intrinsic to the spirit of libraries. Training such as this provides an opportunity to be part of a meaningful conversation around the challenges and opportunities our changing society offers.

References

Groarke, Sarah (2019) Attracting and Retaining International Higher Education Students: Ireland. Dublin: ESRI. https://www.esri.ie/publications/attracting-and-retaining-international-higher-education-students-ireland (accessed 17 August 2020)

Higher Education Authority (2018) Key Facts and Figures: Higher Education 2017/2018. https://hea.ie/assets/uploads/2019/01/Higher-Education-Authority-Key-Facts-Figures-2017-18.pdf (accessed 17 August 2020)

HE4u2 (2017) Creating Intercultural Learning Environments: Guidelines for Staff within Higher Education Institutions. http://he4u2.eucen.eu/wp-content/uploads/2016/04/HE4u2_D2-5_Guidelines_JS_FINAL-2.pdf (accessed 17 August 2020)

HEA (2017) National Plan for Equity of Access to Higher Education 2015-2019 https://hea.ie/assets/uploads/2017/06/National-Plan-for-Equity-of-Access-to-Higher-Education-2015-2019.pdf (accessed 17 August 2020) https://researchrepository.ucd.ie/handle/10197/3607 (accessed 17 August 2020)

Mellon, Bernadette, Cullen, Marie, Fallon, Helen (2013) Implementing an online training course in disability awareness for frontline staff - Experiences at National University of Ireland Maynooth. Sconul Focus, 58. pp. 27-31. http://mural.maynoothuniversity.ie/4607/ (accessed 17 August 2020)

Mestre. L. (2010) Librarians Working with Diverse Populations: What impact does Cultural Competency Training have on their efforts? *Journal of Academic Librarianship,* 36(6), pp. 479-488

UUK, NUS (2019) Black, Asian and Minority ethnic student attainment at UK Universities: Closing the gap https://universitiesuk.ac.uk/policy-and-analysis/reports/Documents/2019/bame-student-attainment-uk-universities-closing-the-gap.pdf (accessed 17 August 2020)

A longer version of this essay was previously published by
An Leabharlann: The Irish Library

PART 2

Poems

Poems from the Maynooth University Library Ken Saro-Wiwa Poetry Competition 2018-2020

Poetic Communities: Judging the Ken Saro-Wiwa Poetry Competition 2018-2020

Jessica Traynor

Although we live in a country known around the world for its rugged mountains, silver lakes and dramatic coastlines, our modern literature and discourse can often feel quite divorced from the natural world. And yet, climate, environment and their impact on humanity are becoming ever more central in international discourse. Even the current Covid-19 crisis is linked to our encroachment on the natural world, allowing new viruses to cross from animals to humans. Although we may look at the natural world through the sterile filter of screens, we can't afford to ignore it.

And so the work of eco-activists such as Ken Saro-Wiwa becomes more important than ever, as does the need to remember and reinforce the passion that led to Irish support for the Ogoni 9 in their struggle to prevent the destruction of their homeland. This vein of passionate activism has been a constant undercurrent in Irish society, and has continued in the Shell to Sea Protests, recent protests around kelp harvesting rights being sold in West Cork, and the rejection of the destructive process of fracking.

The history of activism and social change is often preserved and carried in the stirring words of activists; from Martin Luther King's 'I have a dream', to the poems and letters of Ken Saro-Wiwa, carefully collected by Sister Majella McCarron, and now preserved by the dedicated staff at Maynooth University Library. Words really do make a difference, and are often the means by which meaningful actions are born. And so, the aim in launching the Ken Saro-Wiwa Poetry Competition in 2018 was to engage a new generation of Irish and new Irish citizens with both poetry and eco-activism. We asked poets to explore their love of nature, the landscape and communities around them, and to consider what it might mean to lose these things. Some of our new Irish communities have already experienced terrible loss in their home countries. We felt these stories had something crucial to teach us about preservation of the natural gifts we have been given here in Ireland, and also about empathy, and what new cultures can offer us. We all have so much to teach each other.

In 2018 and 2019, the competition focused on students in transition year in Maynooth Community College. We had wonderful entries, published here, which reflected the students' insightful understanding of the looming climate crisis and how it might shape their futures, and the devastating experiences of those who had had to leave family and friends behind to begin a new life in Ireland. To introduce the concept of the competition and to help the students generate a few ideas they might develop for competition entry, I led some lively workshops in 2018 with students from the local school, in Maynooth University Library. The students visited the Ken Saro-Wiwa Archive, and were really struck by the tangible nature of the artefacts; how they made his story come alive. In the prize's inaugural year, 2018, the winner was Conor Walsh and second prize was awarded to Zofia Terzyk, and at the time I wrote of their poems:

"Conor Walsh's *1859* tackles global pollution through the lens of Edwin Drake's 1859 oil-strike. The poet homes in on the source of so many of our current problems and creates a foreboding atmosphere, foreshadowing the trouble that the reader knows will come. The voice of the oil-worker who slowly discovers the ruin their discovery will wreak on humanity is insightfully evoked. This poem demonstrates ambition, and a lively imagination. A very impressive poem, and a well-deserved first prize."

Sister Majella McCarron, Conor Walsh, Maeve Byrnes, Zofia Teryzk

"Zofia Terzyk is a poet who has a great sense of music, who creates beautiful sounds and images to convey their important message. The image of fists punching metal is timeless and speaks volumes, evoking great industrial waste lands. The title of her poem, *Elements of Life*, is also strong, reminding us of all of the small but important components that make our existence on earth possible. Zofia is a poet with great potential, and I hope she continues to develop her talent in the future."

In 2019, of the prizewinning poems I wrote:

"Jay Vergara's poem *Humans Greatest Sin*, awarded first prize, is an ambitious piece that manages to encompass a large and complex narrative. A child addresses its legacy by interrogating its parents about what humans have done to the planet – a topical dialogue which reflects the anger and anxiety felt by many young people. The imagery of the extinct animals is moving, evoking a universal sense of loss. The rhyme scheme creates movement and momentum throughout."

Elizabeth Akinwande was highly commended for her prose poem *The Bight of Biafra*. This is an intensely lyrical piece, rich with imagery and

Sister Majella McCarron, Jay Vergara, Mercedes Vergara, Jessica Traynor

symbolism. The experience of a child cast out on the waves, surrounded by an oil slick, is evoked with skill. The complex history of the region is called forth through the juxtaposition of gleaming gold and dark, greasy oil. The writer demonstrates a natural talent for world and atmosphere building, and a keenly observant eye.

Shola Akinwande, Elizabeth Akinwande, Sister Majella McCarron

In 2018 and 2019, we feted the winners at the Ken Saro-Wiwa conference in Maynooth University, with readings from the young poets themselves, a talk from Sister Majella McCarron, and contributions from a number of speakers on ongoing eco-activism both in Ireland and in the Niger Delta.

In 2020, to celebrate the 25th anniversary of Ken Saro-Wiwa's death, the competition was expanded further, seeking adult entries nationwide, as well as entries from local schools. As the run up to the competition took place during lockdown, we decided to run two workshops for interested entrants via Zoom, exploring the poems and writings of Ken Saro-Wiwa, alongside a number of other contemporary poems which engage with the environment and activism. The resulting workshops felt like a lifeline in the midst of what had become a lonely and anxious time for many people, and their online nature gave us the opportunity to engage with a wider audience, some of whom may not have been able to travel to Kildare for the workshop. We were delighted with both the quality and quantity of the entries received, and with the insightful engagement with the competition's themes. Of the winning poem, by Lind Grant-Oyeye, I said:

"*African Refugee* is a vibrant, lyrical piece touched by myth. It deals with the changes brought about to ritual by the arrival in a new country, but also with the stories we bring with us. An Ireland gifted with these new stories can only prosper."

Of Eilish Fisher's poem, a close second place, I wrote: "*Night Feedings* tackles the separation of parents from their children on the American border, a barbaric practice. Here, the poet reaches out to these children while holding their own child safely in their arms. It's a study in the transformative power of empathy."

The school entries were also particularly strong this year, bringing together a sense of anxiety at the virus which is currently threatening our way of life, and a passionate will for positive change. The 2020 specially commended and shortlisted poems are also published here, alongside the winning poems from 2018-2020. On reading this year's entries I was delighted to see such a broad spectrum of cultural experiences reflected, so many clever and often unexpected engagements with questions of environment, and such passion for human rights issues.

This year Conor Walsh, winner in 2018, was again awarded first prize in the school category. The poem *Bystander* starts with the wonderful sense image of a cold key in a man's pocket – an image that resonates throughout the poem. This poem interrogates the choices we make every day which impact negatively on others, and how systemic injustice makes these damaging choices easy. A deeply insightful poem.

Second prize was awarded to Ceri Arnott. Her poem *Baking Banana Bread* is an important record of the current moment, as seen through the eyes of a young person. It weaves a subtle metaphor around learned actions and reactions that interrogates how we unknowingly copy the generations that preceded ours. The final image of the bananas with their 'inverted smiles' leaves us in a place of unease.

The shortlisted poems were also very memorable, with *Anticipating* by Christeen Udokamma Obasi imagining the equality an extinction event could bring about. *Know Where*, also by Christeen, explores a place of safety where 'abstract nouns are actual people', and Marykate Donohoe's *Amends* shares an apocalyptic vision of a world shaped by war.

For the past three years, I have read all entries blind, as is the norm in poetry competitions. For me as a judge, the excitement of discovering who the writers are is second only to the anticipation of reading the entries for the first time. Looking back over three years of entries to the Ken Saro-Wiwa poetry competition, I am heartened to see both familiar names and new; names with Irish origins, names from all around the world, and new hybridized names which reflect the joyful coming together of two cultures. In these pages, we have had the rare and precious opportunity to create a space for a new poetic community of likeminded people. This is a legacy of which anyone would be proud. On re-reading and reflecting on these poems, I'm filled with a sense of hope. If these writers and community members can continue to nurture the passion we find in these poems, the future is surely bright.

2020 Entries

First Prize

African refugee

Lind Grant-Oyeye

Broda,
do you feel the frost in this village square,
the way it makes our dance stop?
It's not just because the electricity is gone,
or that the sun forgot it remains a star, we stopped.

O, broda,
are your fingers cold too, the way everything else is
the moon gravitates towards, in its nightly romance feel,
after summer decided to die on us,
steady, slow, then suddenly, puff
the way damfoo buses on Lagos streets puff for breath.

Sista,
do you remember
that man who lost his kaftan while searching
for his father's lost shirt
and his mother's jewels bestowed
upon her by her ancestors?
How he searches.

O, broda,
The bulrushes on our path
now pretend to be carnations,
after they have been stripped
of spring colors
and the climate remains frosty
in its heat, as everything changes.

O, Sista,
who is my broda?

glossary; broda and sista, African vernacular for brother and sister

Second Prize

Night Feedings

Eilish Fisher

For the children separated from their parents at the U.S.-Mexican border, detained in custody and neglected, 2018-present.

I hear you call into the night's fluorescence,
a cage unable to stifle the sound
that travels around the world to this bedside radio.

I move in maternal mindfulness towards a source –
my child's cries puffing like smoke through muted cot bars.
I lift him while your sobs cling and sway.

I stir like tendrils of seaweed in these waves, a pulling of you to me.
I would hold you all if I could-rock you into stony caverns
of peaceful sleep and quiet listenings.

You would hear the tick-tock of the clock on the wall,
rustle of soft-pawed foxes in the woods, the purring
as night's dark harnesses the bee-balm moon.

These days the numbing salve is washed away
as we rock and cry and ache into the long night's waiting.

Special Mention/ Highly Commended Poems

The Prisoner

Caroline Bracken

i.m Ken Saro-Wiwa

Give it up for the prison guard
who breaks ranks to bring me forbidden rations
a piece of bread or cake, a mango
to sweeten my day.
Give it up for the prison guard
who whispers news from my eight brothers
soon to be martyrs but who would rather
be just fathers.
Give it up for the prison guard
who smuggles letters in a breadbasket
crumpled missives from my wife
son or sister.
Give it up for the prison guard
who prays for my soul before it goes
and memorises my words
to pass on.
Give it up for the prison guard
who unlocks my leg irons
and rubs into my wounds
a mother-made poultice.

The Misrecognition of Bodies as Thorns (4 Parts)

Chiamaka Enyi-Amadi

1

– What does it mean to love your neighbor as you love yourself?

when your beauty is stolen from your skin, your hair and the name your mother gave you when she took you from the wet hands of the midwife pale as you were your eyes were still darker than the color of sun-baked mud after the raining-season and she knitted her Africa into your name tighter than the way God knitted it into your jungle hair and safari skin and

now you cannot look yourself in
the eye.

You write yourself love notes in
Morse code.

Each indent is a
question:

Why faces with skin the color of
burnt

caramel or
coal

never grace the crystal
ball,

why love is a word said only with
your back

to the
mirror

and why there is so much
sting

in the silent
bleed

at the end of each
question.

2

– What does it mean to love your neighbor as you love yourself?

when images of tendrils in test tubes are thrown into your living room and they tell you each cell must be stripped from its host for security purposes. There is much talk of contamination but then you read somewhere else...somewhere...on a smaller more transparent screen, that it was babies not tendrils on your flat screen... in incubators and those bits of charred green flesh on glass-covered floors are bodies, are babies, are bloodied overexposed flesh peeling in Eastern heat, not strong enough for the outside world, not strong enough for the once-crisp hospital air, not strong enough for even their mother's touch but

it's been weeks
and all,

there is only rot and dust
now.

Tendrils (or
not)

it is as good a reason to
break

from The Ritual.
Midnight-Media-Rounds.

The daily staring contest with the
digital abyss,

and mourn as best
you can.

They were no family of
yours.

So you will mourn, not
the loss

of shared memories, but the
injustice

(with no thought of
Politics

or other man-made
things).

Simply that they were not
given

the chance to
grow
into an ally or
enemy.

Nipped at the tender bud of
terror.

They will never be old
enough

to feel like victims or
victors,

to feel ostracized or socially
submerged,

to feel that they can be both Arab and
Muslim

without being a threat to Western
security –

to feel the warmth of flesh, not
metal thorns,

to feel, without
pain.

3

– What does it mean to love your neighbor as you love yourself?

if every sunrise doesn't come singin' you can get out of bed today! but maybe you shouldn't since each morning is a constant visceral assault of mortality and you blend into the night and lullabies ring out like elegies or sirens

Because maybe you won't
make it

through the night or passed that
checkpoint

if you're
caught

out. Howling your fear and pride at
the moon!

And there is death and
danger

in the face of
everything
that moves at
you

and
breathes
deep

and
heavy

so that
there is

no air
left

for
your

heaving
lungs.

4

– What does it mean to love your neighbor as you love yourself?

Is it a slow blossoming like a delayed
blessing?

Or is it to become an
injury?

A hurt like the gaping
ground

of an expanding
fault-line?

Settle into
devastation.

Christian your soul
Victim.

Or try

to be

whole again.

Dry Season 1995

Gillian Muir

See-through steam rose up wending its way from the oil drum
Some barefoot women were charcoal boiling palm nuts
Bending, adjusting their wrappers
Cupping out the dark blood liquid into plastic bowls
Colourful against the sooty cylinders.

Okoli was at his farm
He looked up and saw the bursts of smoke fumes
Pop across the sky
He looked down and in came Ken dangling from a string
Ken Saro-Wiwa, son of Ogoniland
Red eyes wiped, the farmer thrust out his machete
There was anger in every swoop as he downed
The overgrown fronds
Spread wide like dead man's fingers
Mocking him.

Beans green shooting earthlings were absent in the undergrowth
Unable to sprout, choked by foreign poison
And strangled at every slick.
Okoli's wives would find nothing to harvest
From their mother ground sick from oil scorch and blood lost.
Nneka arrived, "Sir, ma say come chop now" and she turned
Crackling twigs underneath.
Okoli followed the house-girl's path past the orange buckets and bowls
Past the shaven-head widows and silently
Saluted his son.

Cassandra Speaks

Nora Nadjarian

As a child I was always inventing damage.
I braided snakes into my hair,
let their tongues lick my neck and ears.
They poured cold venom, made my hearing mad.
My fingernails scratched the air,
trying to catch the birds falling from nowhere.
I plucked their feathers and buried their bodies
 in the back garden.

I've foreseen and seen the truth.
I did not mean it, I say, my whole life.
I tear snakes' tongues from their roots,
count mutilated birds.
The scene repeats itself day after day
and every night I dream of falling.

6 Autumns

Owodunni Ola Mustapha

On a warm autumn afternoon 6 years ago,
With an uncertain plan in place
I put on my red flats that serves all my "Israeli Waka" and
Head for the big house on Lower Mount Street
In Dublin 2

I stand by the door
Wondering what to do next
A man in blue beckons to me
And I slowly walk to the heavy sliding door
made of what seems like a guard against recalcitrant trespassers
How may I help you he asks?

I need a place to sleep
I respond in a shaky voice
The two tiny overzealous humans
Standing beside me, oblivious of their surroundings
Ask the man in the blue......
Are you a Policeman?
He giggles, No lads
I work as a security guard here

Put your bag on the table for me
I set my invalid *Ghana Must Go Bag* on the table
He puts something in that looks like ammo from Power Rangers
Beep! Beep!! Beep!!!
He motions for me to walk through a metal door
Welcome to a new world
I mutter under my breath

I will be out of this system in 6 months
I try to convince myself each passing day
The cool breeze and a ray of sunshine blowing across my face
Like the fresh air from Lough Lannagh
Smiles at my broken face and says
6 Autumns is come and gone my lady
Let us wait for the 7th one

Shortlisted Poems

I Will Not Be Shamed

Chiamaka Enyi-Amadi

The women and girls who have suffered female genital mutilation, who live their lives in constant recovery, weighed down by despair and depression, under threat of disease and death, who face shame and stigma in their communities despite the injustices that have been done to them.

We speak too silently
of women who are
broken.
We talk in whispers of
girls whose bodies are
in need of repair.

In rebuilding this
temple, I will not be
shamed.
Let me into the
tabernacle. I will not
be shamed.

I have washed my sheets
and cleaned my linen of all
past sins. I will not be
shamed.

Let me in. I want to build a god
in my own image. I will not be
shamed.

I want to carve an idol of
desolation. I will not be
shamed.

In the name of a love that is
self-reflective,

in the name of a god that will
watch me (dis)honor my body and
will not condemn my soul, I will
not be shamed.

To not be judged according to the
weight of that first Sin in the
Garden of Eden. I will not be
shamed.
To be a body in
exile. I will not
be shamed.

To be a body in
recovery. I will not
be shamed.

Let me in

To be free of
Resentment

To be gracefully
un-transcendent

To be reborn in the hot mouth of
temptation

To be worthy in
imperfection. I will not
be shamed.

In this aging
Impermanence, I will
not be shamed.

Let me in

I want to build a god
that will not
transcend me in this
tragedy of being.

I will not be shamed
for loving this
body. I will not
be shamed.

Tlaloc

David Butler

Tired of burnings, bulldozers, charred lungs,
Tlaloc the Rain God decamps from Mayan
rainforests, rides the bloated trade winds,
comes to reign over the Old World.
Days on end the swollen earth has
swallowed till it's soft as blotting-paper.
The sun is an aspirin dissolving
in a gauze of soaked cotton.
Storm-drains clog, rivers turn reptilian,
shed their alluvial taint in basements,
swell to bursting, silver-plate the floodplains.
Tlaloc moves over the face of the waters,
looks on desolation wrought by man,
calls up the other Weather Gods.

White Rose

David Butler

i.m Sophie Scholl, executed 1943

too easy, to doubt
    a scruple's chance
in a world in flames
one petal
    quivering
in a thistle-field
one
  out-of-step
    footfall
in a fall of jackboots
or
a lone voice
calling
from the wolf's jaws:
'such a splendid
    sunny
        day
and I have to go'

The Disappearing

Bairbre Flood

Yul Brynner's bald head is on the wall opposite,
an advert for used cars:
'Yul never beat our offers'.

You take a drag from your cigarette,
the slight dawn light unflattering.
A green tint on the windowsill.

George Wassouf is playing low on your phone
and you're suddenly reminded
of the street you used to live on.

Maybe they'll rebuild it someday, I say.

You tut –
not like our tuts –
but the Arabic tut which means no,

Will they rebuild my neighbours
from their graves
you say.

Aethon of Thessaly

Jimmy O'Connell

Borrowed from Ovid

And Gaea stood on the precipice
Of time and watched in helpless
Anguish as Aethon impiously took
His axe and cut down her sacred oak.

In righteous anger she travelled
To the sun-seared plains and requested
Void to enter the stomach of Aethon
As he slept in his palace of plenty;

He awoke in a rage of hunger demanding
He be satiated by every variety of food
His servants could prepare and place
Before him, but the more he ate the more

His hunger was fed like fuel to fire;
And so, still unfulfilled, he quit his domain
And entered the woodlands of the earth,
Tore down the trees and rended the very

Bark that protected them and snapped off
Its branches with his teeth and chewed upon
The sweet leaf on each limp twig. His thirst
Was such that he stood by the banks of the river

And supped its sweet waters, but it was never
Enough; he longed to taste the salt in each drop
Of the whale-bathing sea and so he drained
The many abyssal oceans. As yet unsatisfied

He fell to his knees before the unploughed fields
And fed upon the various clays from where
The seeds of future plants were stored,
As well as the plethora of worms, insects

And micro-grubs that crept and crawled there;
Aethon then beheld the sheep and cattle
That grazed in the meadowed and fallowed
Fields and, in his frenzy of mindless hunger,

Tore into their flesh and gnawed their bones
Until their butchered carcasses were as nothing;
And even to his daughter he turned who,
Frightened and paralysed with fear and torment,

Could not escape his cavernous greed;
And at last he stood upon the barren earth
Where lay before him a landscape of shale,
Stones, wind-shifted sands and layer upon layer

Of grim grey rock; seeking yet more and more
To inebriate his craw-gapped need his left hand
Raised itself in front of his bloated face and he
Salivated as each finger quivered in fear and dread

And sunk his teeth into them. Gaea fell
To her knees and watched as Aethon,
In his last fit of uncontrollable greed,
Devoured himself.

A d e j e (Ah-dee-jay)

Liam O'Neill

Her name was Mercy, and she sat
on a chair beside me at the factory.
At first, she hardly said anything to
her fellow workers except her given
name 'Mercy', and that she was
born in a small village called *Adeje*
situated near in the Nigerian delta.

Seeking relief from the boredom
of the assembly line, we spoke of
the places we came from in Ireland.
Warming to the openness of the
factory workers, Mercy spoke of
her home village and the members
of her family still residing there.

There was an accident she said,
a great unforgettable catastrophe.
She was busy in a kitchen, cooking,
when she heard an explosion, and
ran outside her front door to see
lines of burned and charred bodies
lying on the roadway to her town.

The day before, poor people came
and drilled holes in the oil pipeline
that ran near the village, and locals
found out and ran to the pipe with
pots and pans to salvage some oil for
themselves. Mercy lost her mother,
her brother and twin sister cousins.

Two hundred and fifty people dead.
Mercy saw neighbours face down
in the river where they had tried to
quench the flames that engulfed them.
Adeje – once her playground became
a place of desolation, and she knew
she had to leave the village for good.

Oil is a curse on a country Mercy says,
it brings corruption and contamination
to everything it touches. Poverty killed
her family and friends; poverty and the
greed of local politicians and European
Oil men in suits. Injustice killed Adeje;
blew all of Mercy's childhood away.

Doctor Osman

Liam O'Neill

At night, in Ireland, I ready myself for bed.
At night, in Sudan, Doctor Osman is being dragged
in to a 'ghost house' to be beaten with sticks.
Doctor Osman spoke too loudly of murders and rapes –
way too loud for the ears of Governmental officials.

In the morning, as I dress for work,
Dr Osman's captors inject acid into his urethra.
He loses consciousness as I pull out of the drive.
Later, when I'm eating my lunch, they throw
Osman into the street and leave him there for dead.
Passers-by pass him by, for fear of similar hospitalities.

In the evening, as I relax and watch a show on TV,
Osman's captors come for him once again, but he escapes
and crawls to a safe place to avoid punishment and death.

As I switch off the TV and lock the door, Osman,
without the aid of anaesthetic and sterilised equipment,
cuts deep with a blade into his skin and bladder,
to release the poisonous liquids bubbling inside.

At night, in Ireland, I ready myself for bed.
At night, in Ireland, Doctor Osman gets ready too.
He escaped the torture in Sudan to endure the distress of
'Direct provision' in my own country.

After a third appeal, the Judge agreed, the Doctor ticked
all the boxes for an official refugee.
'People are being tortured every day', Doctor Osman says
to any reporters who will listen,
'No-one speaks up for them', he adds, 'they need a voice.'

At night, in Ireland, I settle down to watch movie.
At night, in Sudan, more innocent people are being dragged
into 'ghost houses' to be injected with poisons.

'They need a voice', Doctor Osman says.

'Who will be their voice?'

A Lamentation of Swans

Glen Wilson

Their song is life in the making,
necks curved under pressure,

in reflection they make a heart
but we ripple it away

with the stale crumbs
that breach the surface

with what we don't need
and what they didn't ask for.

They've been chased from the fens of history,
given ponds and then we vilify

the females for rumours
of apocryphal arms broken,

the males blamed for the lusts of gods,
or those who style themselves as such

all based on the accepted lies
of nights so far removed.

It is their untamed ignorance
we envy, for it can't be earned

for we learn as Leda did
those in power take

what they can never have.

for the house of the planter is known by the trees

– Austin Clarke

Joseph Woods

Sweet suburbs of Harare
and their bungalows that are known
by the trees, parades of purple
jacarandas, poinsettias and flame trees
and beyond their boundaries,

swards of green edging to the road
with sprinklers levitating even in dry season.
And the sweeping, swish sweeping
by liveried gardeners always in blue

who bid 'baas' or 'sir'
to my passing 'good morning'
or who pause and chat among
themselves in front of compounds
whose walls are topped

with sometimes sizzling electric
wires and unforgiving electric gates
that snap shut after a glimpse
of the interior. Every avenue
looks the same and were it not

for the colour of ornamental trees
you could get lost among the reassuringly
foreign street names, Sandringham, Churchill and Windsor...

Longlisted Poems

Gigo

Yewande Adebowale

belligerent origins of waste ablaze in bonfire
spewing eternal flames of flared gas from a draconian chimney
rising to the heavens in soot and particles
provoking acidic rainfall, yes mother earth gives gigo with reckless abandon

plastic dumpling bubbles litter the seas in plethora
ocean cum dumpsite of non-biodegradable decommissioning
oil barrels disposed offshore drop after drop refusing to cease
aquatic life depleted by milestones and yardsticks

intense heat waves in their stride have come
voluminous floods fill the trenches breaking down floodgates
windbreaks now desertified
you receive what you give

the earth cries uncontrollably refusing to be comforted
the seas rage without fail displaying utmost displeasure
you receive what you give
garbage in, garbage out

Before Love Was Legal

Gavin Bourke

Widowed,
by a living
and loving husband.
An impossible double life,
a pregnant wife
and a child under the age
of five.

Living in two worlds
at once.
Loyal to neither,
not by choice either.
Married in their twenties
without any anxieties.

Glamorous photos
hiding true preferences,
blinkered love and ambitions.
Picture frames
still perfectly intact,
long after the breaking
of a sacred pact.

Children of the Revolution

Mary Melvin Geoghegan

for Waad al-Kateab

Her daughter Sama (Arabic for sky)
was born in Aleppo, in a hospital
founded by her father –
The paediatrician who delivered her
the only doctor available
was killed four months after the birth.
Her father Hamza in just twenty days
carried out 890 operations and
cared for over 6,000 wounded people.

In a close-up, lingering intimacy
the sadness of hospital staff
barely aware of the camera rolling.

Through the Window

Mary Melvin Geoghegan

A rook lands on a roof out there
wings braking and feet outstretched.
A blackbird already perched stands its ground
and now a V of geese flies north-west.
To count how many bird silhouettes
Jan Van Eyck could have painted
in this span of skyline –
While, the world still flows outwards
and spreads, percolating into countless
particulars.

The day they drove old Dixie down

Lind Grant-Oyeye

I wish my mother could see me now,
and all she thought would ruin it all
"driving in cars with boys"
their youth deep in the seat pocket of random
car seats that do not live to tell the stories of fleeting romance

I wish my pa could see me now,
the way he saw them politicians speak random, from afar-
a language meant for only those who know how to drive blind,
the way muddy waters find their way to mud ditches

I wish I could talk about flowers
and pollination
or even something as random as illustrative
colors of fairy tales

You see, old Dixie and all the talk
of the boys here and there,
are stuck, where it all began,
paying tribute
to the branches of the Niger river,

waiting for a language gods and humans would understand,
while they pluck happy melons on the streets of Africa
and continue godding*, after the harvest.

**playing god*

A Journey Through Santa Elena **Paul McCarrick**

It is hot.

Breaths have been taken, a drink of water.
There is a fear to surviving the night.

Do I need to know how to swim? I say.
He finishes his drink, goes up a gear and takes an exit.
No. Not really. You don't need to know how to swim,

as long as you have the will to live.

I have that will for life to live more than ever before now,
bring me home, run me back to stand in the world's waters,
renew myself again, rush me from this deep closing heat.

We travel through valleys, walls carved out from ruptures,
themselves blessed by being untouched by measurements,
centimetres, man-made precision. Hitting one side we travel
on solid and different ground. Taking a step in, we journey

through ravines being born, through rockslides and gorges
and come out again, fresh, warm, fine in the cool rays
of Mexico's sun. We could track the trails of coyotes,
or snakes, or see in the tv blue sky the ghosts of Camacho chiefs,

looking down at us new invaders, or men not on horseback,
but standing watch, sighting us in their cold vision of steel
moving within the shadow, breathing in the wind,
crying in the river, lying on the earth. It is hot.

We deep dive past borders made from hand and land and pass
to the point of being born again, baptised by mother earth and
her holy clean water, wholesome, to stand again and feel the sun
and the wind and this earth on this land, anew.

Give me the question again,
and I will say I have it now,
I have it now and so much more.

It's Your Word Against Theirs

Nora Nadjarian

1.
She says no again and some words are not accepted,
fall on the floor like her clothes.
Who told you to wear
the almost nothing of a whore?
The hands on her mouth and a deaf scream and her heart
beats like a loudspeaker blasting it all over town.

2.
The blood she lost, she cups it in her palms
and smears her body with a blanket of shame.
Her limbs are dead, her insides deciding to die.
The doctor is suspicious of her mouth which opens
and closes without a sound. This silence must be operated on.
What is left of me?

3.
She speaks in contradictions, an unreliable witness.
Madwoman. Can't remember who, how many,
exactly at which point.
Whatever you say will be recorded. Used against you.
Questions pace up and down in the tiny room,
words roll out of her mouth like broken teeth.

4.
She comes out drenched and shivering, hiding her face
from cameras. Shock stands next to her and she asks
for a loving word, one word that loves her, to hear
at least one word of reply from the world. How easy it was
to enter the neon lights and sway to the music.
The volume gradually goes up to deafening.

Smell of The Fish

Philomena Obasi

Sleeping so calm
Arising so sluggish
Aches and pains
On torn raffia mat!

Moonlight…
Cricket sounds, fish smoke
Kinsmen in circles
Grandma smoking fish
Few, but many mouths

Sharing, fighting
Kings share
Community together

Our bare, natural existence
All for One…One for All

God's Plan – Ken Saro-Wiwa's Spiritual Tale

Patrick O'Siochru

My child, what do you see?
Stars, many of them.
My child, what do you feel?
Pain. With intention to heal.
My child, what do you smell?
Happiness. A smell of hope.

You must know, in a time to come, I must leave,
But do not forget, we share this land, from which we breathe.
In all our might, and in all our will,
They will never destroy this beauty, nor shall we in return lustre to kill.

These lands, they speak in whispers,
The darkness is our guide,
The light is our bearer.
Let it not be man, who rules this country,
May you to feel no more pain against a monstrous adversary.
The spirit of what is right,
This will determine your faith, command this evil, and be sure to put up a good fight.

My child, I must go, my time is up,
As the sands command it, for time does not move slow,
For which time will let us meet, separate us at last upon this false acquit.
Do not look so frail, my child,
You have your mother's eyes.
If you feel lost, look up to the skies,
For God sees all, for no quest is too small, or too great,
And I, your watchful eye, looking down carefully from St. Peter's Gate.

School Category 2020

First Prize

Bystander

Conor Walsh

As he reached into his pocket,
He felt the cold metal key.
It sends a shiver up his spine,
But that will dissipate in an instant.

He turns around to take one last look at the site
Before it becomes a profit driven retail outlet.
He sees the family,
A father, mother and four children,
Attempting to hold back the emotions.

As they pack frantically
He ponders how they must feel,
The emotions that are consuming them.
They certainly won't simply, dissipate.
A tear comes to his eye.

But he reminds himself,
This is not his fault.
He is only doing his job,
He's an innocent bystander.

What more could he have done?

Second Prize

Baking Banana Bread

Ceri Arnott

The recipe lay for all to see.
I leaped up and beamed at Mommy
As she yanked the ripest yellow banana from its friends.

She explained: its fellow bananas understood.
They knew their friend would provide a world of good
For us humans.

The golden dome rose for miles.
To help I reached up and turned the dials
Just like Mommy did.

Alas by the ding the poor bread was burned.
When I ate it my stomach churned.
Perhaps I should not have copied Mommy.

Of course, now I understand,
Such high temperatures are impossible to withstand.
The crust had been penetrated, the potential surpassed.

To this day, I remember the rest of the bunch –
Those haunted inverted smiles,
Mourning, abandoned, unripe.

Shortlisted Poems

Amends

Marykate Donohoe

Lost are the soldiers, in the mountainous sky.
Absent are their morals,
As the blood of truth runs dry.

The blood shall run, and nourish this earth.
From the seeds of selfless sacrifice,
The acts of change shall bloom.

Anticipating

Christeen Obasi

When that day comes
The D-day
When we start to contemplate on our reasons for existing
Our privilege
Then and only then would be so similar
All races and species alike
We will bow our heads and accept our fate
Like slaves
We will regret our injustice and unfairness
Then, we will be judged
By something greater than
Our unfiltered tongues
And piercing eyes.

Know Where

Christeen Obasi

This is my refuge – here
This is where I can think, relax, laugh happily
And be
No one judges me here, I'm satisfied here
Abstract nouns are actual people here
Feelings and emotions likewise
This is the land of the candid
Or so I believe
Or so I want to believe
At least, I'm happy here
And that's all that matters.

School Category 2019

First Prize

Humans Greatest Sin

Jay Vergara

Mother! Father! My little child said,
As she rushed into my bedroom and leaped onto my bed
With a look of excitement and wonder
"What are these creatures?" she pondered
Images of creatures, which existed long ago
Rushed into extinction from the damage humans did so
We invaded their homes, and took their resources
If they fought back our way we would force in
We've famished our Earth, our greatest sin
"But Mother, Father surely we humans have done great!
"Humans are Prefect" said in Christian Faith
A lie it is, it's far from the truth
Does the destruction we've done make us absolute?
Not a single patch of nature seen in our cities
Our plan of action is only feeling pity
Our Buildings and factories standing in nature's way
Nowhere to expand, left to decline and fade away
The air we respirate, is corrupted and twisted
There was once a time it wasn't as misted
Now with every breath, feels like you're choking
Without a cigarette, you're basically smoking
Disappear if we would, life would thrive
Stay if we would, life would die
With the look of excitement no longer on her face
She bitterly said "What is wrong with the Human Race?"

Second Prize

The Bight of Biafra

Elizabeth Akinwande

Freezing.
Wrapped up in myself, the sea rocks violently almost like the tempest raging on in my head, shaking my security. This boat is ripped apart by the waves and my life by the war.

Nigeria torn in two like the weak (silk) cotton it was made out of.
The needle and thread in their hands.
Instead of sowing the fabric of our country back together: they pricked us, bonded us and left us no choice but to leave.

The Bight of Biafra.
The oil on which the boat rocks is the violence in which my mind absorbs.
Constantly replaying in my memory, chipping off pieces of me as it repeats.

Biafra, Nigeria, Torn, Silk, Black Gold; Dust.

I don't know how many days have passed, maybe 12, maybe 33.
Too lost in my own world to notice. Too lost in my old world to notice.
Trying to savor the last morsels of the boy that I'd become. So exposed to the terror of the man I would be and where this oil slick would take me.

The heaving and labored breaths of a woman stole my attention.
Her sickly green tainted the mahogany of her skin and she heaved once more over the side of the boat. Sweat permeated from her tattered clothes.
I knew her name was Nneka. She told me warmly when we first got onto the boat, her matronly smile was a source of comfort for the first couple of days.
She spoke Igbo just like I.

And she told me "ihe niile ga-adi mma nwam". (Everything will be fine my son)

I reached my hand to her shoulder and mustered all the comfort I had out of my sad heart into hers. I whispered "Anyi ga adi nma"(we will be okay) and she closed into me and wept.
She wept for the both of us and everything we had lost, everything we had grown to love and everything in between.

School Category 2018

First Prize

1859

Conor Walsh

Standing there at the side of the drill,
Watching the oil, black as coal,
Being poured into the container,
Seeing Mr. Drake's face as he watched on,
the pride clearly evident,
It was simply magnificent.

Thinking of what heroes we were,
How the world will
Change because of us,
The legacy we will
leave behind.

We rushed to tell them what we had found,
How different their reaction was to
What I had envisaged.
We were not glorified heroes,
We were monsters.

They told us "if used this oil would be the death of the human race."
Sitting here now knowing that we almost damaged the earth in an
 unthinkable fashion,
I am comforted in the knowledge that we will always remember the
 dark day
When oil tried to kill us.

We must take the lessons learned from that day and allow them to
influence Everything we do,
The mistakes of that day must
Never be made again.
Because if they are the Consequences will be grave.

Second prize

Elements of Life

Zofia Terzyk

Wounding words pollute the water,
thrown away dreams damage the earth,
tears of lost love blow out fires,
sharp knives cut the trees,
furious fists punch the metal,

Only heroes can save us now,
fight the crime and turn back time,
Peace, Love, Hope and Joy,
cleanse the world,
bring back Life,

So strong yet so shatterable.

Third prize

The Sun Shines Down

Eva Paturyan

The sun shines down
On views changed for the worst.

Yesterday it shone on a paradise of trees,
Today it shone on a barren wasteland.
Last week, on waves of water,
This week, on waves of plastic.
Last year, on faces lit with hope,
This year, on faces fearing for their future.
Last century, on wildlife left untouched,
This century, on wildlife fast disappearing.

The sun shines down on this earth we call home,
And will continue to shine while we undo our wrongs,
Until it shines down once again
On views changed for the better.

Highly Commended

Pollution

Áine Dooley

A global problem arises,
Like large plumes of smoke
From the industrial chimneys of greed-driven goliath
Causing the earth and air to choke

A global problem arises,
Like large mounds of plastic waste
Accumulating on our oceans floors and surfaces
Condemning all sea-life encased

A global problem arises,
Like large volumes of environmental noise
Emitted from unwanted human activities
So loud, so continuous, it annoys

A global problem arises,
Like air, sea and noise pollution
Merging to form climate change
A growing concern that requires your eco-hero contribution

Melt

Maeve Byrnes

Melting cap of ice,
We turn the earth to an oven,
Sets the earth on fire
Cooking the scorched lands,
Killing the beauty of Earth.
Our Earth, too precious.

Five Poems by Sister Majella McCarron (OLA)

Dying Village

No piercing siren to rise you,
No flaring light to guide you,
No nearby neighbour's love to reach you,
Dying village.

No caring friend to conceal you,
No towering soldier to defend you,
No rushing firemen to quench you,
Dying village.

No speeding ambulance to ferry you,
No humble priest to bury you,
No Red Cross pennant to fly for you,
Dying village.

No urgent phone to ring for you,
No loud-pitched radio to plead for you,
News is blocked in fear of you,
Dying village.

No strong one comes to hold you,
As children are torn from you,
A stranger's voice to wail for you,
Dying village.

Dawn comes late for you,
Vultures chuckle over you,
Our deepest human shame is you,
Dying village.

Too few prophets spoke for you,
Years of scribes and pharisees denied you,
Evil powers abandoned and beggared you,
Dying village.

The world turned its back on you,
May God himself be good to you,
And hope renew in you,
Dying village.

1995

A Night of Death 10 – 11 – 1995

Large sobs of rain
Gusts of pain
Flowed along the lane
Of Ken Saro-Wiwa Park.

Umbrella shroud
Unrolled
To cradle
Your precious name.

I'd brought you home
Given you a space
A place to be mourned
Until freed to your own.

You were waked in Ireland
On a street of shame
Renamed
And washed
With Ireland's rain

1995

Letters

(to Ken Saro-Wiwa in detention)

1

I stand
On the edge of space –
Thrusting thoughts
To tested hands
Carrying on
In silent song
To yonder place!

2

I make
Boat to sail –
With words
That bind
Heart to life
Where horror's
Rife!

3

I weave
Threaded thoughts
That trace and lace
Soul to soul –
Fragile chords
That hold and hold
Greed's relentless roll!

4

I pray
That hope should hold
Wind blow open
The hidden fold –
No story of mine
Remain untold
Abandoned to bitter cold.

1995

Warmth in the Wind

Harsh wounded lands-
Stumbling in the storm
I stand and look
In horror
At what's gone on!

From way out there
A stranger comes
To hold my hand
In muted unison-
I understand
What he's done!

A quiet voice
Soothing a weeping heart
Slowly calms the harm
Quells the fear
Dries the tears
For what's gone on and done!

1996

Ajido

Afternoon –
Hot, heavy heat!
Stillness
Seeping everywhere,
As lethargy creeps along human limbs
Constraining busy agitated restless me
Into
The space of myself
Drawn by my physical line.
Releasing feelings
And emotions
And memories
Of pure integration ...
Bowed down and held by the head
Together in one wholeness!
The long narrow pin strip
Winding its way
Under sheltering trees
To the silence of Ajido
With its shapeless space
Held by absence
Of pavements!

1995

Developing a world view through a diversity of voices: A Reflection

David Rinehart

The poems in this book give us the unique opportunity to explore the thoughts and imagination of people from a diversity of backgrounds and experiences. Some were born and raised in Ireland; some have come to live in Ireland from different countries including Nigeria and America, while others currently live in other countries around the world. Many of the poets share a common sentiment of how tired, distressed and enraged they are by the vast array of social problems smattered across our globe. These voices resonated with me, reminding me of stories I heard in the United States, my home country. The global North and the global South share an unjust and lopsided history, yet we, those of us reading this book, those of us who are activists, those of us who are poets and writers and academics, we are coming together to demand justice, equality, and to save our planet.

I moved to Ireland in August of 2018. It was then, through the Maynooth University Ken Saro-Wiwa archive, that I learned the story of Ken Saro-Wiwa and the Ogoni 9 which shed light on egregious acts of violence and injustice in Nigeria. With a background in Latin American Studies, parts of this story felt familiar to me and drew me in. Now, working in Special Collections & Archives at Maynooth University Library, I'm in an incredibly privileged position to have a small part in this story, to contribute my very own words, experience, and knowledge. For this, I am very grateful.

The essays and poems in this book shed light on the many malignant ways in which capitalism and colonialism have wreaked havoc on our world. These forces destroy our planet, shape and sharpen racial tension and violence, widen the wealth gap, fuel sexism, encourage xenophobia and anti-immigration sentiment, and the list grows on. Some of the poets in the School Category in this collection describe a doomed earth where drilling for oil, the overproduction of plastic, and the politics that keep these dirty practices moving are destroying our planet, and subsequently their dreams. The anxiety these students feel, which is also evident in some of the poems in the adult category, is tangible and readers will be able to relate to this. It is but another consequence of global hypercapitalism and colonialism. These symptoms of capitalism are felt here as much as anywhere else. Take the Americas as an example.

In the Americas, the modern imperial force known as the United States of America has manipulated elections, staged coups, and allowed unimaginable violence to sweep the region so that foreign interests, namely American corporations, can stake their claims to land and benefit from cheap labour. These regimes smack down labour unions, tear apart labour laws, and take away the rich and fertile land from its own people. For example, in the 1950's, Jacobo Árbenz became the second democratically elected president in the more than 100 years of Guatemala's independence. He ran on a platform of agrarian reform. Árbenz wanted to take the unused land owned by foreign corporations, namely the United Fruit Company, and give it back to the Guatemalan people. The United Fruit Company had majorly powerful stockholders in the US government, such as the Dulles Brothers, John and Allen, who were respectively the Secretary of State and Director of the CIA. The company was concerned about Árbenz ideas and his popularity. The CIA staged a successful coup d'état.[1]

This has been followed by decades of violence, poverty, coups, and corrupt governing. One of the largest groups of immigrants undergoing a dangerous journey and seeking asylum in the United States are Guatemalans. The United States is doing everything it can to prevent entry and those who do get in, work hard long days for less than the minimum wage, which is already far less than a living wage. When I learned about the Ogoni people, a story of corporate greed and personal wealth at the cost of others' health, wellbeing, peace, and land use, it reminded me of the story of Guatemala and the United Fruit Company.

Whether it is Guatemala or Nigeria, or any other countries, the story is similar enough for us to recognise a global trend: a trend in which imperialism has transformed into global capitalism. We see the so-called "developed world" profiting from the labour and land of the so-called "underdeveloped" world. The global North reaping the benefits and raping the land of the global South. We see people fleeing their homes, not because they want to, but because they have no other choice. Liam O'Neill's poem, *Doctor Osman*, describes something I am sure many of us have felt at some moment:

1 Schlesinger, S., & Kinzer, S. (2005). *Bitter fruit: The story of the American coup in Guatemala.* Cambridge (Massachusetts): Harvard University, David Rockefeller center for Latin American Studies.

At night, in Ireland, I settle down to watch a movie.
At night, in Sudan, more innocent people are being dragged
into 'ghost houses' to be injected with poisons.

The same powers taking advantage of these countries, are putting up borders and illegalizing and marginalizing the very people most affected by their actions.

Eilish Fisher, in her work *Night Feedings*, exposes the monstrous, centuries old genocidal and racist practices implemented by the United States of dehumanizing 'others.' In this case, those dehumanized are migrants from the global South. These practices are so ugly that they go as far as to separate parents and children from each other, keeping children in cages, like animals. To give context to these practices: when the United States established what is referred to as the Prevention Through Deterrence policy back in 1994, namely putting up a physical border and increasing the amount of Border Patrol agents, it was acknowledged that it would lead to migrant deaths at the hands of the cruel desert climate. They called these deaths unavoidable 'collateral damage.'

Since 1998, over seven thousand migrants have died in the desert according the U.S. Border Patrols own statistics.[2] This does not include the thousands of migrants who have been declared missing. To add fuel to the fire, private prison corporations such as CoreCivic, inc. (formerly known as CCA), make billions of dollars a year for holding migrants in their euphemistically named 'detention centers.' They make 160 dollars per person per day, giving them incentive to overcrowd and to lobby for stricter border policies.[3]

Those of us who recognize this violence and oppression need to be heard: to remain silent is to be complicit. That is what I understand Ken Saro-Wiwa to mean by Silence is Treason, or the trope we hear throughout the global Black Lives Matters movement, Silence is Complicity. With that in mind, I would like to reflect on Ken Saro-Wiwa's words,

As we subscribe to the sub-normal and accept double standards, as we lie and cheat openly, as we protect injustice and oppression, we empty our

2 Southwest Border Deaths by Fiscal Year. (2019). Retrieved September 2, 2020, from https://web.archive.org/web/20190109174045/https://www.cbp.gov/sites/default/files/assets/documents/2017-Dec/BP Southwest Border Sector Deaths FY1998-FY2017.pd

3 Rinehart, D. C. (2018). *Walking the Fine Line: Legal Precarity Along the U.S.-Mexico Border* (Unpublished master's thesis). University of Florida.

classrooms, denigrate our hospitals, fill our stomachs with hunger and elect to make ourselves the slaves of those who ascribe to higher standards.[4]

Reading the entries to the Ken Saro-Wiwa Poetry Competition strengthened my sense of solidarity with oppressed people everywhere. Some of the poets write about migration. Some reflect on homes and lands left behind, such as Lind Grant-Oyeye in *African Refugee*, or what their bodily experience is in this world, which we feel viscerally in Chiamaka Enyi-Amadi's *The Misrecognition of Bodies as Thorns*. Some of these poets, like me, have been raised and socialized by the global North and are waking from a hazy dream to see this western world for what it is, an ugly dragon hoarding gold and hurting those who come near.

These poems give us the most empathetic tool imaginable, to view the world through the eyes of others. You'll hear the beauty of humanity, people who enjoy entertainment, who struggle to raise their children, who love infinitely, who strive, who feel, who cry and laugh. We can smell, taste and feel what the poet smells, tastes and feels, like the smell of "Grandma smoking fish," in Philomena Obasi's poem *Smell of the Fish*. In just a few lines, we can be someone else, experience as someone else, understand someone else, and, from that, we become a better more complete person. We get a taste of humanity and what it is as a whole. We nurture and grow our empathy. This should ignite a desire within us to keep reading, to keep learning, to keep listening. Poetry is a tool to look outward at the world and also to look inward, to understand others and the effect of our actions as a global community on others.

I cannot speak for those of you whom I consider myself an ally, what I can do is tell you that I am here, and I am listening. I will follow your lead and learn from you only when and if you want to teach me. I will not pressure you or ask you, I will simply keep my eyes and ears open for those of you who are speaking and writing and painting and playing. I have learned so much from you and I have so much more to learn. The best lesson I have learned, as of yet, is that being anti-racist, an ally, a feminist, open minded and loving is a process that lasts our whole lives. It's a way of living, not an accomplishment.

These poems, the work of Ken Saro-Wiwa and the Ogoni 9, and the work of Maynooth University Library in keeping Ken Saro-Wiwa's work

4 Dickson, A. (2005, November 10). Against forgetting. Retrieved September 02, 2020, from https://www.theguardian.com/culture/culturevultureblog/2005/nov/10/againstforgett

and words alive, has helped shape me and helped guide me on this journey. The voices in these poems and essays have helped me develop my understanding of the world I am fighting for. These words show us what the world is, what the world should and should not be and how we can work together to achieve justice and equality.

I welcome the publication of this volume which highlights the work of Trócaire and other people and organisations to fight for a more just world. The story of Ken Saro-Wiwa's struggle for justice for the Ogoni people is shocking yet also profoundly mobilising. Trócaire campaigned strongly for the release of the Ogoni 9 and we were shocked when we heard on 10th November 1995 that all nine had been executed. Trócaire continues to campaign for justice in the face of state and corporate violations of human rights. The writings of Ken Saro-Wiwa and those inspired by his work keep the flame of justice lit.

— **Dr Caoimhe De Barra**,
Chief Executive Officer, Trócaire

Books of this nature, well written and researched, contribute to struggles such as that of the Ogoni. They ensure that peaceful protests continue to echo and that the attempt to censor our struggle by the execution of the Ogoni 9 by Abacha and Shell does not succeed. This resonates with the agitations of the present day and the need to liberate minds towards fulfilling dreams for which the nation and our communities were built. The publication of *I am a Man of Peace: Writings Inspired by the Maynooth University Ken Saro-Wiwa Collection* highlights the shift towards peaceful protests as a means to a more sustainable end. It is a justification that one's dreams for the greater good can still be attained without bloodshed. Ken preached, lived, and practiced peace in his search for justice even in the face of stiff opposition. Years after his death, Ken's ideologies and philosophies have proved relevant today. My thanks to Sister Majella and the Maynooth University Library for keeping the story alive, again through this collection.

— **Dr Owens Wiwa**, Executive Vice-President,
Regional Director of West and Central Africa
and the Country Director in Nigeria for the
Clinton Health Access Initiative (CHAI)

Sister Majella McCarron's choice of Maynooth University for this unique donation was particularly appropriate, given the University's long involvement with issues of inclusion and justice in Ireland and abroad. This deep-rooted commitment is today articulated in our University Strategic Plan, where a strategic goal is *'to build on our achievements to date and become a model University for equality, diversity, inclusion and inter-culturalism, where social justice, addressing inequality and empowering people are central to our mission.'*

This book makes a significant contribution to that goal.

— **Dr Gemma Irvine**, Vice-President of Equality & Diversity,
Maynooth University

Writing can urge us to pause, to think, and to discover what we really want to say. Sharing that writing calls for courage and support. In this publication, as Jessica Traynor suggests, we see the evidence of a new generation of Irish and new Irish citizens engaging with complex issues through poetry. Reading their work, we may begin to understand more about each other and ourselves. By bringing together these Irish and new Irish voices the book contributes to building a shared discourse which is essential for trust, community and hope.

— **Dr Alison Farrell**, Founder of the Summer Writing
Institute For Teachers (SWIFT) and Co-founder,
Irish Network for the Enhancement of Writing (INEW)